Invisible Waves:
The Complete History and Future of Radio

Douglas Donovan

ISBN: 979-8-3495-5960-0

First Edition 2025

Part I: The Foundation Years (1865-1920)

Chapter 1: The Invisible Force - Maxwell's Mathematical Prophecy

In the sweltering heat of a London summer in 1865, as Queen Victoria's Britain stood at the pinnacle of its industrial might and the distant echoes of the American Civil War's conclusion reached European shores, a quiet revolution was about to unfold in the pages of scientific literature. On June 16, 1865, a manuscript bearing the modest title "A Dynamical Theory of the Electromagnetic Field" was delivered to the printing house of Taylor and Francis, destined for publication in the prestigious Philosophical Transactions of the Royal Society. The author, James Clerk Maxwell, a soft-spoken Scottish physicist with penetrating eyes and an unruly beard, had no way of knowing that his mathematical equations would one day enable voices to travel across oceans, images to dance through the air, and humanity to reach toward distant stars with invisible waves.

What Maxwell had accomplished in those densely packed pages would prove to be nothing short of revolutionary. He had not merely unified two seemingly disparate forces of nature—electricity and magnetism—but had revealed the mathematical foundation for an entire spectrum of invisible waves that would transform human civilization. Among these predicted waves were the radio frequencies that would, within a few decades, shrink the world and connect humanity in ways that even the most visionary minds of 1865 could scarcely imagine.

The Inheritor of Faraday's Vision

Maxwell's journey to this momentous discovery had begun a decade earlier, in 1855, when he first encountered the experimental work of Michael Faraday, the brilliant but largely self-taught English experimenter who had revolutionized the understanding of electricity and magnetism. Faraday, working in the basement laboratory of the Royal Institution, had discovered the mysterious "lines of force" that seemed to emanate from magnets and surround electric charges. With iron filings scattered around a magnet, he could make these invisible forces visible, revealing elegant curved patterns that suggested some deeper underlying reality.

But Faraday, despite his extraordinary experimental intuition, lacked the mathematical sophistication to express his insights in the precise language that theoretical physics demanded. His descriptions of electromagnetic phenomena were qualitative and pictorial—he could see and describe the effects, but he couldn't reduce them to mathematical relationships that would allow precise predictions and calculations.

Maxwell, blessed with both exceptional mathematical ability and deep physical intuition, immediately recognized the profound importance of Faraday's experimental observations. In 1855, at the age of just 24, he presented his first paper on the subject, "On Faraday's Lines of Force," to the Cambridge Philosophical Society. This initial work represented Maxwell's first attempt to translate Faraday's qualitative insights into mathematical language, but it was merely the beginning of what would become a ten-year intellectual odyssey.

Over the course of the next decade, from 1855 to 1865, Maxwell embarked on an increasingly ambitious project to not only mathematize Faraday's discoveries but to construct a complete theoretical framework that would encompass all electromagnetic phenomena. This was no mere academic exercise—Maxwell was attempting to discern the fundamental laws that governed some of nature's most mysterious forces.

The Mechanical Analogies and Growing Understanding

In his early work, Maxwell relied heavily on mechanical analogies to understand electromagnetic phenomena. The Victorian mind was thoroughly mechanical in its outlook—if something couldn't be explained through gears, pulleys, springs, and rotating wheels, it seemed somehow less real. Maxwell initially imagined the electromagnetic field as a kind of elaborate mechanical system, complete with rotating vortices and idle wheels that could account for the various electromagnetic effects that Faraday had observed.

In his 1861-1862 paper "On Physical Lines of Force," Maxwell developed these mechanical analogies in extraordinary detail. He envisioned a medium filled with molecular vortices that could rotate around magnetic field lines, with small particles acting as "idle wheels" between adjacent vortices to prevent them from rotating in the same direction. This elaborate mechanical model allowed him to derive mathematical relationships between electric and magnetic phenomena, and it was in this work that he first introduced the crucial concept of "displacement current".

But even as Maxwell developed these mechanical models, he began to suspect that they were merely stepping stones to a deeper understanding. The mechanical analogies were useful for visualization and mathematical development, but they felt increasingly artificial and contrived as the theory grew more sophisticated.

The Revolutionary Leap of 1865

By 1865, Maxwell was ready to make a conceptual leap that would fundamentally transform physics. In his paper "A Dynamical Theory of the Electromagnetic Field," he abandoned the mechanical analogies that had guided his earlier work and focused instead on the electromagnetic field itself as the fundamental reality. This represented a dramatic shift in thinking—rather than trying to explain electromagnetic phenomena through mechanical models, Maxwell was proposing that the field was a real physical entity that existed independently of any underlying mechanical structure.

The mathematical framework that Maxwell developed in this paper was both elegant and comprehensive. He reduced all known electromagnetic phenomena to a set of twenty field equations in twenty variables, creating for the first time a unified mathematical description of electricity and magnetism. But Maxwell's equations did far more than simply organize existing knowledge—they revealed hidden truths about the nature of electromagnetic phenomena that no one had suspected.

The crucial breakthrough came when Maxwell considered how insulators store electrical energy and made a small but profound correction to Ampère's Law. Ampère's original law described how magnetic fields are created around electric currents, but it had a subtle flaw when applied to situations involving capacitors—devices that store electrical charge on separated metal plates. When electric charge flows onto the plates of a capacitor, the current appears to stop at the gap between the plates, creating a mathematical inconsistency in Ampère's Law.

Maxwell resolved this problem by proposing that something he called "displacement current" flows through the insulating material (or even empty space) between the capacitor plates. This displacement current wasn't the flow of actual electric charges, but rather represented the changing electric field in the region between the plates. From a purely mathematical standpoint, this correction was necessary to make Ampère's Law consistent with other known electromagnetic relationships.

The Astonishing Discovery

But when Maxwell worked through the mathematical implications of his corrected equations, he discovered something that took his breath away. The electromagnetic fields he was describing didn't simply exist as static entities—they formed a medium capable of supporting waves that could propagate through space. Even more remarkable, when he calculated the speed at which these waves would travel, the result was approximately 310,740,000 meters per second—a number that was virtually identical to the recently measured speed of light.

This numerical coincidence was far too precise to be accidental. Maxwell, with characteristic scientific restraint, wrote in his 1865 paper: "The agreement of the results seem s to show that light and magnetism are affections of the same substance, and that light is an electromagnetic disturbance propagated through the field according to electromagnetic laws".

This seemingly modest statement represented one of the greatest unifications in the history of science. Maxwell had demonstrated that light—that ethereal phenomenon that had puzzled natural philosophers since ancient times—was simply oscillating electric and magnetic fields propagating through space. The mystery of light had been solved, and in solving it, Maxwell had revealed something even more profound about the nature of reality itself.

The Prediction of Invisible Waves

Maxwell's equations revealed that light was just one small part of a vast electromagnetic spectrum. If visible light consisted of electromagnetic waves with wavelengths of roughly half a micrometer, then there should exist electromagnetic waves with much longer wavelengths and much shorter wavelengths. The equations made no distinction between different wavelengths—they predicted that all electromagnetic waves, regardless of their frequency or wavelength, should behave according to the same fundamental principles.

This meant that somewhere in nature there should exist invisible electromagnetic waves with wavelengths measured in meters, kilometers, or even longer. These waves should be capable of being generated by oscillating electric charges and detected by their effects on other electric charges. Maxwell had not only explained the nature of light but had also predicted the existence of an entire spectrum of invisible waves waiting to be discovered and harnessed.

The implications were staggering. If such waves could be generated and detected, they might provide a means of communication that required no wires, no physical connection between sender and receiver. Messages could be transmitted through the apparently empty void of space, carried by invisible waves that obeyed the same mathematical laws that governed light itself.

The Mathematical Foundation for a Wireless World

Maxwell's field equations were more than just a theoretical achievement—they were a blueprint for technologies that wouldn't be realized for decades. Hidden within those elegant mathematical relationships were the principles that would enable the development of radio, television, radar, satellite communication, and countless other applications that depend on the generation, propagation, and detection of electromagnetic waves.

The equations showed that accelerating electric charges would radiate electromagnetic waves into space. They revealed how these waves would propagate through various media and how they would interact with matter. They predicted how electromagnetic waves could be focused, reflected, and manipulated. In essence, Maxwell had provided humanity with the theoretical foundation for the entire field of wireless communication.

But theory alone was not enough. Maxwell's equations predicted the existence of radio waves, but they didn't show how to generate them efficiently or detect them reliably. The transition from mathematical theory to practical technology would require the genius of numerous inventors and engineers working over the next several decades.

The Unfinished Symphony

Maxwell himself seemed to grasp the profound implications of his discovery, but he would not live to see its practical realization. In the years following the publication of his 1865 paper, he continued to refine and develop his electromagnetic theory, eventually reducing his twenty field equations to the four elegant equations that bear his name today. He also made fundamental contributions to other areas of physics, including kinetic theory and statistical mechanics.

Tragically, Maxwell died of stomach cancer in 1879 at the age of just 48, having lived only fourteen years beyond his great electromagnetic breakthrough. He died without ever seeing a radio wave generated or detected, without knowing that his equations would one day enable human voices to span the globe, without realizing that his mathematical prophecy would transform civilization itself.

But Maxwell's equations lived on, carrying within them the seeds of a technological revolution. In laboratories across Europe and America, a new generation of experimenters began to test the predictions embedded in Maxwell's mathematical framework. The age of wireless communication was about to begin, built upon the foundation that Maxwell had laid with his profound insight into the electromagnetic nature of light.

The Prophet's Legacy

Freeman Dyson, one of the most distinguished physicists of the twentieth century, would later call Maxwell's 1865 paper "the most important event of the nineteenth century in the history of the physical sciences". This assessment, delivered with the benefit of more than a century of hindsight, captures the revolutionary nature of Maxwell's achievement. In a single theoretical framework, he had not only unified electricity, magnetism, and light but had also opened the door to technologies that would define the modern world.

The path from Maxwell's mathematical prophecy to practical wireless communication would prove to be long and challenging. Heinrich Hertz would spend years in his laboratory trying to generate and detect the waves that Maxwell had predicted. Guglielmo Marconi would struggle to transform laboratory demonstrations into practical communication systems. Countless other inventors, engineers, and scientists would contribute to the development of radio technology.

But it all began in 1865 with a Scottish physicist, his mathematical equations, and a remarkable insight about the fundamental nature of electromagnetic phenomena. Maxwell had revealed that the universe was permeated with invisible waves obeying precise mathematical laws. He had shown that these waves could, in principle, be harnessed to carry information across vast distances without any physical medium other than space itself.

The implications extended far beyond practical communication. Maxwell's electromagnetic field theory represented a fundamental shift in how physicists understood the nature of reality. Instead of explaining all phenomena through direct mechanical interactions between particles, Maxwell had shown that fields themselves were real physical entities that could exist independently in space. This conceptual revolution would prove crucial for the development of relativity theory and quantum mechanics in the twentieth century.

The Dawn of the Electromagnetic Age

As the nineteenth century drew to a close, Maxwell's equations stood as a monument to the power of mathematical physics to reveal hidden truths about nature. The equations predicted not only the existence of radio waves but also X-rays, gamma rays, infrared radiation, and ultraviolet light. They showed how electromagnetic waves could be polarized, interfered with one another, and diffracted around obstacles. They provided the theoretical foundation for understanding how electromagnetic radiation interacts with matter.

Most importantly for the future of radio, Maxwell's equations revealed the fundamental principle that would make wireless communication possible: oscillating electric charges radiate electromagnetic waves into space, and these waves can induce oscillations in distant electric charges. This reciprocal relationship between motion of charges and electromagnetic waves would prove to be the key to both transmitting and receiving radio signals.

The stage was now set for the next act in the drama of wireless communication. Maxwell had provided the theoretical foundation, but someone would need to demonstrate that his predicted waves actually existed. Someone would need to show that they could be generated, controlled, and detected. Someone would need to transform Maxwell's mathematical prophecy into technological reality.

That someone would be Heinrich Hertz, working in his modest laboratory in Karlsruhe, Germany, in the 1880s. But Hertz's experiments, brilliant as they were, would represent only the first tentative steps toward practical wireless communication. The full realization of Maxwell's vision would require the efforts of many brilliant minds working over several decades, each building upon the mathematical foundation that Maxwell had established in 1865.

The invisible force that Maxwell had described mathematically would soon become visible through its effects, audible through its ability to carry sound, and indispensable to human civilization. The age of radio was about to begin, made possible by the profound insights of a Scottish physicist who had glimpsed the mathematical structure underlying one of nature's most fundamental phenomena. Maxwell's prophecy was complete; now it remained for others to fulfill it.

Chapter 2: Sparks Across the Void - Hertz Proves the Theory

The winter of 1886 was unusually harsh in Karlsruhe, Germany, where snow lay thick on the cobblestone streets and coal smoke hung heavy in the frigid air. Inside the physics laboratory of the Karlsruhe Institute of Technology, a young professor named Heinrich Rudolf Hertz was conducting experiments that would soon make history. At twenty-nine years old, Hertz possessed the rare combination of theoretical brilliance and experimental skill that marked the greatest physicists of his era. What he was about to accomplish in that modest laboratory would transform Maxwell's mathematical prophecy into tangible reality and open the door to the age of wireless communication.

Twenty-one years had passed since Maxwell published his revolutionary electromagnetic field theory, yet no one had successfully demonstrated that the predicted electromagnetic waves actually existed. The scientific community had largely accepted Maxwell's equations on theoretical grounds—they were mathematically elegant and explained known electromagnetic phenomena with remarkable precision. But science demands more than mathematical beauty; it requires experimental validation. The invisible waves that Maxwell had predicted remained exactly that: invisible, undetected, and existing only in the realm of mathematical possibility.

Hertz was about to change all that. In the autumn of 1886, while experimenting with a pair of Riess spirals in his laboratory, he noticed something that would redirect the entire course of his career and, ultimately, the trajectory of human civilization. When he discharged a Leyden jar into one of the coils, a spark appeared in the other coil, even though the two were not physically connected. This spark across the void suggested that some form of energy was traveling through space from one coil to the other—exactly the kind of phenomenon that Maxwell's theory predicted.

The Reluctant Pioneer

Hertz's path to this momentous discovery had been anything but direct. Seven years earlier, in 1879, his mentor Hermann von Helmholtz had suggested that Hertz focus his doctoral dissertation on testing Maxwell's electromagnetic theory. Helmholtz, one of the most respected physicists in Germany, had even proposed a "Berlin Prize" problem at the Prussian Academy of Sciences for anyone who could experimentally prove electromagnetic effects in the polarization and depolarization of insulators—something that Maxwell's theory predicted but no one had yet demonstrated.

Despite Helmholtz's confidence that Hertz was the most likely candidate to win the prize, the young physicist initially declined the challenge. He couldn't envision how to build an apparatus capable of testing Maxwell's predictions experimentally. The theoretical framework was elegant, but the practical challenges of generating and detecting electromagnetic waves seemed insurmountable. Instead, Hertz chose to work on the more conventional topic of electromagnetic induction, a phenomenon that was well understood and experimentally accessible.

But Hertz did produce an analysis of Maxwell's equations during his time at Kiel, showing they had more validity than the then-prevalent "action at a distance" theories. This theoretical work deepened his appreciation for Maxwell's insights and prepared him for the experimental breakthrough that would come later. Sometimes the most profound discoveries emerge not from direct pursuit but from the convergence of preparation, opportunity, and an observant mind.

The Apparatus That Changed the World

By late 1886, Hertz had received his professorship at Karlsruhe and was experimenting with various electrical phenomena when he observed the mysterious spark jumping between his Riess spirals. Suddenly, he realized he had found "an idea on how to build an apparatus" that could tackle Helmholtz's Berlin Prize problem, even though the actual prize had expired uncollected in 1882. The spark across the void suggested a way to generate and detect the electromagnetic waves that Maxwell had predicted.

Working with characteristic German precision and thoroughness, Hertz designed and constructed an experimental apparatus that was both elegant and effective. His transmitter consisted of a dipole antenna made from two collinear one-meter copper wires with a spark gap between their inner ends and zinc spheres attached to the outer ends for capacitance. This arrangement served as what we would now recognize as a half-wave dipole antenna—a configuration that remains fundamental to radio engineering today.

To power his transmitter, Hertz used a Ruhmkorff coil, an early type of transformer that could generate high-voltage pulses of about 30 kilovolts. When these high-voltage pulses were applied across the spark gap, they created violent electrical discharges that caused the charges on the antenna wires to oscillate rapidly back and forth. According to Maxwell's theory, these oscillating charges should radiate electromagnetic waves into the surrounding space.

For his receiver, Hertz constructed a resonant single-loop antenna with a micrometer spark gap between its ends. This device was positioned at various distances from the transmitter to detect any electromagnetic waves that might be propagating through space. The genius of this arrangement lay in its simplicity: if electromagnetic waves from the transmitter reached the receiver, they should induce oscillating currents in the loop antenna, and these currents should be strong enough to produce sparks across the receiver's spark gap.

The Moment of Truth

On that historic day in late 1887, when Hertz first activated his spark-gap transmitter and observed an answering spark in his distant receiver, he witnessed something that no human being had ever seen before: the direct evidence of Maxwell's predicted electromagnetic waves. The receiver was positioned as far as 1.5 meters away from the transmitter, with no physical connection between them, yet sparks danced across its gap in perfect synchronization with the transmission.

This was not merely an electrical phenomenon confined to wires and circuits—it was proof that energy could travel through apparently empty space in the form of invisible waves. Hertz had generated what we now call radio waves in the very high frequency range and had demonstrated that they could be transmitted and received across a significant distance. The implications were staggering, though Hertz himself would later admit that he saw no practical applications for his discovery.

The sparks that jumped across Hertz's receiver gap were caused by the arrival of electromagnetic waves from the transmitter, which generated violent electrical vibrations in the receiving antenna. These vibrations were Maxwell's predicted electromagnetic fields made manifest—oscillating electric and magnetic fields propagating through space at the speed of light, carrying energy from the transmitter to the receiver without any material medium.

The Systematic Validation

Having achieved this initial breakthrough, Hertz embarked on a systematic program of experiments designed to verify every aspect of Maxwell's electromagnetic theory. Between 1886 and 1889, he conducted a series of increasingly sophisticated experiments that would definitively prove that the effects he was observing were indeed the electromagnetic waves that Maxwell had predicted.

Starting in November 1887 with his paper "On Electromagnetic Effects Produced by Electrical Disturbances in Insulators," Hertz began sending a series of reports to Helmholtz at the Berlin Academy. These papers documented his methodical investigation of electromagnetic wave properties, including crucial experiments in 1888 that showed transverse free-space electromagnetic waves traveling at a finite speed over measurable distances.

In one of his most ingenious experiments, Hertz positioned his oscillator about 12 meters from a zinc reflecting plate to produce standing waves. Each complete wavelength measured about 4 meters long—a scale that made it possible to map the wave structure with his ring detector. By carefully moving his receiver antenna through the region between the transmitter and the reflector, he could record how the wave's magnitude and component direction varied with position.

This standing wave experiment allowed Hertz to demonstrate several crucial properties of electromagnetic waves. He located the nodes (points of minimum signal strength) and antinodes (points of maximum signal strength) that formed when the transmitted waves interfered with their reflections from the zinc plate. The distance between adjacent nodes revealed the wavelength of the electromagnetic radiation, while the pattern of nodes and antinodes provided direct evidence that these were indeed waves propagating through space.

Hertz found that the length from one node to an antinode to another node equaled one complete wavelength. He could identify a node by detecting a spot where no spark occurred in his receiving antenna's spark gap, then move the antenna systematically until he found regions where sparks did occur. This methodical mapping of the wave structure provided unambiguous evidence that electromagnetic energy was propagating in the form of waves with well-defined wavelengths and frequencies.

Measuring the Speed of Light

Perhaps most remarkably, Hertz was able to calculate the frequency of his electromagnetic waves and demonstrate conclusively that they traveled at the speed of light, exactly as Maxwell's theory predicted. By measuring the wavelength through his standing wave experiments and knowing the frequency of oscillation in his transmitter circuit, he could apply the fundamental relationship that the speed of a wave equals its frequency multiplied by its wavelength.

The result was stunning: Hertz's electromagnetic waves traveled at approximately 300,000 kilometers per second, the same speed that had been measured for light. This numerical agreement provided powerful confirmation that light and radio waves were indeed different manifestations of the same underlying phenomenon—electromagnetic radiation. Maxwell's great unification was no longer just a theoretical proposition but an experimentally verified fact.

Hertz also demonstrated that his electromagnetic waves could be reflected, refracted, and made to produce interference patterns, just as light waves could. He showed that the waves exhibited polarization—their electric field oscillations had a definite orientation in space—and that this polarization could be controlled and manipulated. All of these properties matched Maxwell's theoretical predictions with remarkable precision.

The Unexpected Discovery

During the course of his radio wave experiments, Hertz made an entirely unexpected discovery that would later prove crucial to the development of quantum mechanics. He observed that when he placed a piece of glass in front of his receiving loop antenna, the size of the spark decreased significantly. When he replaced the glass with a quartz plate, which allows ultraviolet light to pass through, the spark returned to its original size.

This phenomenon—now known as the photoelectric effect—puzzled Hertz completely. As he commented at the time: "the effect is striking and yet totally puzzling. Naturally it would be nicer if it were less puzzling; however, there is some hope that, when this puzzle is solved, more new facts will be clarified than if it were easy to solve".

Hertz had no way of knowing that he had stumbled upon one of the key experimental observations that would later lead Albert Einstein to propose the quantum nature of light. The photoelectric effect would eventually earn Einstein the 1921 Nobel Prize in Physics and help establish quantum mechanics as one of the foundational theories of modern physics. It was a fitting irony that Hertz, in the process of confirming Maxwell's classical electromagnetic theory, had also uncovered evidence for the quantum nature of light that would eventually supersede classical physics in many domains.

The Validation Complete

By 1889, Hertz had completed his systematic validation of Maxwell's electromagnetic theory. He had not only proved that electromagnetic waves existed but had also demonstrated that they possessed all the properties that Maxwell's equations predicted: they traveled at the speed of light, they could be reflected and refracted, they exhibited interference and polarization, and they carried energy through space without requiring any material medium.

These experiments established definitively that light and radio waves were both forms of electromagnetic radiation obeying Maxwell's equations. The theoretical framework that Maxwell had constructed through pure mathematical reasoning had been confirmed in every detail by Hertz's careful experimental work. The invisible force that Maxwell had described mathematically was now visible through its effects, measurable through its properties, and controllable through human ingenuity.

Hertz's work revealed that the electromagnetic spectrum extended far beyond the narrow range of wavelengths visible to the human eye. His radio waves had wavelengths measured in meters, while visible light had wavelengths measured in tiny fractions of micrometers. This suggested that there might exist an entire spectrum of electromagnetic radiation with wavelengths ranging from the extremely short to the extremely long—a prediction that would be confirmed by the subsequent discovery of X-rays, gamma rays, infrared radiation, and other forms of electromagnetic radiation.

The Ironic Blindness to Applications

Despite the revolutionary nature of his discoveries, Hertz himself failed to recognize their practical implications. His pursuit of electromagnetic waves had been motivated purely by scientific curiosity and his desire to test Maxwell's theoretical predictions. He never imagined that radio waves would have any practical purpose beyond advancing human understanding of natural phenomena. When asked about the potential applications of his work, Hertz reportedly replied that his waves were "of no use whatsoever."

This blindness to practical applications was characteristic of pure scientific research in the nineteenth century. Scientists like Hertz were motivated primarily by the desire to understand nature's fundamental principles rather than to develop new technologies. The transition from scientific discovery to practical application typically required the efforts of inventors and engineers who could envision how theoretical insights might be transformed into useful devices.

Yet Hertz's "useless" discovery would prove to be one of the most practically significant scientific breakthroughs in human history. Within a few years, other investigators would build upon his work to develop practical wireless communication systems. Guglielmo Marconi, in particular, would recognize the commercial potential of Hertzian waves and would dedicate his career to transforming Hertz's laboratory demonstrations into a global communication network.

The Legacy of Sparks

Hertz's experimental confirmation of Maxwell's electromagnetic theory marked a turning point in the history of physics and technology. For the first time, invisible waves had been generated, controlled, and detected by human devices. The gap between theoretical prediction and experimental reality had been bridged, opening the way for a cascade of technological developments that would transform human civilization.

The spark gaps that Hertz used in his transmitter and receiver would evolve into increasingly sophisticated devices for generating and detecting radio waves. His dipole antenna design would become the foundation for radio engineering, while his techniques for measuring wavelength and frequency would establish the methodology for radio science. Even the unit of frequency—the hertz—would be named in his honor, ensuring that his contribution to electromagnetic science would be remembered as long as humans work with wave phenomena.

Perhaps most importantly, Hertz had demonstrated that the seemingly abstract mathematical relationships in Maxwell's equations corresponded to real, measurable phenomena in the physical world. This validation of Maxwell's theoretical framework established electromagnetic field theory as one of the pillars of classical physics and provided the foundation for countless technological developments, from radio and television to radar and satellite communication.

The Bridge Between Worlds

Hertz's experiments represented a bridge between two worlds: the world of pure mathematical theory that Maxwell had explored and the world of practical technology that would soon emerge. His spark-gap transmitters and loop antennas were primitive by modern standards, but they embodied the fundamental principles that would guide the development of all subsequent radio technology.

The electromagnetic waves that danced between Hertz's apparatus in that Karlsruhe laboratory were the same waves that would soon carry news across oceans, entertainment into homes, and eventually enable global communication networks that would shrink the world and connect humanity in ways that even the most visionary minds of the 1880s could not have imagined.

When Hertz died tragically young in 1894 at the age of just 36, he had lived to see the initial confirmation of his discoveries by other researchers but had not witnessed the explosion of wireless technology that would follow. His brief career had been sufficient to transform Maxwell's mathematical prophecy into experimental reality and to establish the scientific foundation for the age of radio communication.

The sparks that had jumped across the void in Hertz's laboratory were more than just electrical discharges—they were harbingers of a new era in human communication. The invisible waves that Maxwell had predicted and Hertz had proven would soon carry the voices, music, and messages that would unite the world in an invisible web of electromagnetic communication. The age of radio was about to begin, built upon the foundation of sparks across the void that Hertz had first observed on that historic day in 1887.

Chapter 3: The Great Inventor Race - Tesla vs. Marconi

In the final decade of the nineteenth century, as the world stood poised on the threshold of a new millennium, two brilliant minds were locked in an epic race that would determine who would be remembered as the father of wireless communication. On one side stood Nikola Tesla, the enigmatic Serbian-American inventor whose wild hair and piercing eyes seemed to crackle with the same electrical energy that flowed through his revolutionary devices. On the other stood Guglielmo Marconi, the young Italian aristocrat whose combination of technical skill, business acumen, and social connections would ultimately secure his place in history as the practical pioneer of radio.

This was more than just a competition between inventors—it was a clash between two fundamentally different approaches to innovation. Tesla, the visionary theorist, conceived of wireless communication as part of a grand unified system that would transform human civilization. Marconi, the pragmatic entrepreneur, focused on solving specific technical problems and building commercially viable communication systems. Their rivalry would span decades, involve massive financial interests, and ultimately require the intervention of the highest court in the United States to resolve.

Yet the story of wireless communication's birth involves more than just these two towering figures. The race to harness Hertz's electromagnetic waves attracted brilliant minds from around the world: Alexander Popov in Russia, Jagadish Chandra Bose in India, Reginald Fessenden in North America, and many others whose contributions have been overshadowed by the Tesla-Marconi rivalry. Understanding this broader context reveals that the invention of radio was not the work of a single genius but rather the culmination of efforts by numerous pioneers, each building upon the discoveries of those who came before.

Tesla's Visionary Leap

The story begins not in 1895 when Marconi conducted his famous backyard experiments, but three years earlier in 1892, when Nikola Tesla was already envisioning a world connected by invisible waves. Tesla had arrived in America just a decade earlier, carrying little more than the clothes on his back and a letter of introduction to Thomas Edison. By 1892, he had already established himself as one of the most innovative electrical engineers in the world, having developed the alternating current induction motor that would revolutionize electric power generation and distribution.

Tesla's approach to wireless communication emerged from his broader vision of electrical transmission. Unlike other inventors who were focused on point-to-point communication, Tesla conceived of the entire Earth as a conductor that could be used to transmit electrical energy and information to any point on the globe. This revolutionary concept went far beyond what others were contemplating—Tesla was imagining a wireless world decades before such ideas seemed remotely practical.

In 1893, Tesla delivered a series of lectures that would prove to be prophetic. Speaking to the Franklin Institute in Philadelphia and later to the National Electric Light Association in St. Louis, Tesla described wireless communication systems in remarkable detail. He explained how electrical oscillations could be transmitted through space and received at distant locations, and he demonstrated devices that embodied these principles. These weren't merely theoretical speculations—Tesla was showing working prototypes that illustrated the fundamental concepts of wireless transmission and reception.

The St. Louis demonstration was particularly significant because it occurred in front of a large audience of electrical engineers and scientists who could witness Tesla's wireless devices in operation. Using high-frequency alternating currents, Tesla illuminated gas-filled tubes without any wire connections, demonstrating that electrical energy could indeed be transmitted through space. He also showed how wireless signals could be used to control mechanical devices—a concept that would later evolve into remote control technology.

The Brilliant Boat Demonstration

Perhaps Tesla's most impressive early demonstration of wireless technology occurred in 1898 at Madison Square Garden in New York City. Before an astonished crowd, Tesla unveiled a radio-controlled robotic boat that he operated remotely using wireless signals. The boat could start, stop, turn left or right, and even turn its lights on and off, all in response to commands transmitted from Tesla's control device. The audience was so amazed by this seemingly magical demonstration that some suspected Tesla of using trained animals or hidden accomplices to operate the boat.

This demonstration was remarkable for several reasons. First, it showed that wireless signals could not only carry information but could also be used to control mechanical devices—anticipating the modern field of robotics and remote control. Second, Tesla's boat incorporated what we would now recognize as the fundamental principles of digital communication, using specific coded signals to trigger different responses. Third, the demonstration proved that wireless communication was not just a laboratory curiosity but could have practical, real-world applications.

Despite the impressive nature of this demonstration, Tesla struggled to find practical applications for his wireless technology. He approached the U.S. Navy with his radio-controlled boat, envisioning its use as an unmanned torpedo or reconnaissance vehicle, but military officials couldn't grasp the potential of wireless-controlled weapons. Tesla's ideas were simply too far ahead of their time, and his somewhat eccentric personality may have made it difficult for conservative military officials to take him seriously.

Tesla's patent number 613,809, filed in 1898, described his "Method of an Apparatus for Controlling Mechanism of Moving Vehicle or Vehicles". This patent covered not only the radio-controlled boat but also the broader principles of wireless remote control. It was a foundational document that anticipated many of the wireless communication technologies that would emerge in the twentieth century.

Marconi's Practical Breakthrough

While Tesla was developing visionary concepts for wireless communication, a young Italian inventor named Guglielmo Marconi was taking a more focused and practical approach to the same challenge. Born into wealth and privilege in 1874, Marconi had access to resources and social connections that would prove crucial to his success. Unlike Tesla, who was driven by grand theoretical visions, Marconi concentrated on solving specific technical problems that would make wireless communication commercially viable.

Marconi's breakthrough came in 1895 when he was just twenty-one years old. Working in the attic of his family's estate at Pontecchio, near Bologna, Italy, Marconi successfully transmitted wireless signals across his family's backyard using equipment that bore striking similarities to the devices Tesla had described in his 1893 lectures. Marconi's transmitter used a spark gap to generate electromagnetic waves, while his receiver employed a coherer—a device invented by French physicist Édouard Branly—to detect the incoming signals.

The coherer was a crucial component that made practical wireless telegraphy possible. This device consisted of metal filings in a glass tube that would cohere (stick together) when subjected to electromagnetic waves, allowing electric current to flow. After each signal was received, the coherer had to be "decohered" by tapping it, which separated the metal filings and prepared the device to receive the next signal. While primitive by modern standards, the coherer was sensitive enough to detect the weak electromagnetic waves that early wireless transmitters could generate.

Marconi's initial success in his family's backyard convinced him that wireless communication had enormous commercial potential. Unlike Tesla, who seemed more interested in pushing the boundaries of what was theoretically possible, Marconi immediately recognized that practical wireless communication could revolutionize maritime transportation, military communications, and news reporting. This business-oriented perspective would prove to be one of Marconi's greatest advantages in the race to develop wireless technology.

The Aristocratic Advantage

Marconi's social background provided him with advantages that Tesla, despite his genius, could never match. As the son of an Italian aristocrat and an Irish heiress, Marconi had access to influential circles in both Italy and Britain. When his early wireless experiments in Italy failed to generate the support he needed, Marconi traveled to London in 1896, where his mother's connections helped him gain introductions to key figures in the British scientific and business communities.

In London, Marconi found a receptive audience for his wireless technology. The British Empire depended heavily on maritime trade and naval power, making reliable ship-to-shore communication a matter of national strategic importance. Marconi's wireless telegraphy system offered the possibility of maintaining contact with ships far from land, a capability that could revolutionize both commercial shipping and naval operations.

Marconi's timing was perfect. The British patent system was well-developed and offered strong protection for inventors, while British investors were experienced in funding new technologies. Perhaps most importantly, the British Admiralty and commercial shipping companies could immediately see practical applications for wireless communication. Within months of his arrival in London, Marconi had secured his first British patent and was conducting demonstrations for government officials and potential investors.

The contrast with Tesla's experience in America was stark. While Tesla was certainly recognized as a brilliant inventor, his eccentric personality and tendency to make grandiose claims about future possibilities sometimes made it difficult for him to secure the focused support he needed for wireless development. Tesla's interests were also divided among many different projects, from alternating current systems to high-voltage experiments, while Marconi concentrated almost exclusively on wireless telegraphy.

The Patent Wars Begin

As both Tesla and Marconi continued to develop their wireless systems, the question of who had priority in the invention of radio became increasingly contentious. Tesla's supporters pointed to his 1893 lectures and his 1898 patent as evidence that he had conceived of wireless communication before Marconi. Marconi's advocates countered that priority should go to the person who first demonstrated practical, working wireless communication systems rather than to the one who first articulated theoretical concepts.

The legal situation was complicated by the fact that Tesla and Marconi had filed patents in different countries and for different aspects of wireless communication. Tesla's patents tended to be broader in scope, covering fundamental principles of wireless transmission and reception. Marconi's patents were more specific, focusing on particular technical solutions to practical problems. This difference in patent strategy would have significant consequences as the wireless industry developed.

In 1900, the United States Patent Office awarded Tesla patent number 645,576 for his "System of Transmission of Electrical Energy," which covered key aspects of wireless communication. This patent described the fundamental principles of tuned transmitters and receivers—the foundation of all practical radio systems. Tesla's patent included the crucial concept of using tuned circuits to achieve selectivity, allowing multiple wireless stations to operate simultaneously without interfering with each other.

Marconi, however, was not content to let Tesla's patent claims go unchallenged. Backed by powerful financial interests, including funding from J.P. Morgan, Marconi's company launched a series of legal challenges designed to establish Marconi as the rightful inventor of radio. These patent disputes would continue for decades, involving some of the most prominent patent attorneys and legal experts of the era.

The Transatlantic Triumph

While the patent battles raged in the courts, Marconi was achieving practical successes that captured the world's imagination and solidified his reputation as the pioneer of wireless communication. His most spectacular achievement came on December 12, 1901, when he successfully transmitted the first transatlantic wireless signal from Poldhu, Cornwall, to St. John's, Newfoundland—a distance of over 3,500 kilometers.

The transatlantic transmission was a remarkable technical achievement that required pushing wireless technology to its limits. Marconi's transmitting station at Poldhu used a massive antenna system consisting of fifty vertical wires suspended from towers over 60 meters tall. The transmitter itself generated unprecedented power levels, using banks of capacitors to store energy and massive spark gaps to create the electromagnetic oscillations needed for long-distance transmission.

On the receiving end in Newfoundland, Marconi used a completely different approach. Instead of a permanent antenna system, he employed a wire antenna lifted high into the air by a kite—a brilliantly simple solution that allowed him to establish a receiving station quickly and economically. The signal he received was faint—just three dots representing the letter "S" in Morse code—but it was unmistakably artificial and had clearly traveled across the Atlantic Ocean.

The success of the transatlantic transmission transformed public perception of wireless communication overnight. Newspapers around the world trumpeted Marconi's achievement, and investors rushed to fund wireless ventures. The practical applications were immediately obvious: ships crossing the Atlantic could maintain contact with land, isolated regions could be connected to the global communication network, and military forces could coordinate operations across vast distances.

Tesla, meanwhile, was pursuing his own ambitious wireless project. Convinced that wireless communication was just one aspect of a broader system for transmitting electrical power, Tesla persuaded the financier J.P. Morgan to invest $150,000 in the construction of a massive transmission tower at Wardenclyffe on Long Island. Tesla's vision for Wardenclyffe went far beyond simple wireless telegraphy—he envisioned a system that could transmit both power and information to any point on Earth.

The Wardenclyffe Dream and Disappointment

Tesla's Wardenclyffe project represented his most ambitious attempt to realize his vision of wireless world communication. The facility was designed around a massive tower topped with a distinctive mushroom-shaped conductor that Tesla believed would enable efficient transmission of electrical energy through both the Earth and the atmosphere. The underground portion of the installation included an extensive network of pipes and conductors designed to couple the transmitter to the Earth's natural electrical properties.

Tesla's concept for Wardenclyffe was decades ahead of its time, anticipating technologies that wouldn't be practical until the development of modern electronics. He envisioned a system that could not only transmit telegraph messages but also telephone conversations, photographs, and even moving pictures—essentially describing what we would now recognize as multimedia broadcasting. Tesla also believed that Wardenclyffe could provide wireless electrical power to remote locations, eliminating the need for power transmission lines.

Unfortunately, Tesla's grand vision proved to be his downfall. As construction of Wardenclyffe progressed, costs spiraled far beyond Tesla's initial estimates. When he approached J.P. Morgan for additional funding, the financier was no longer interested in supporting what appeared to be an increasingly impractical project. Morgan was also aware that Marconi's more focused approach to wireless telegraphy was already generating commercial success and patent protection.

Tesla's inability to complete Wardenclyffe was a devastating blow to his wireless ambitions. Without additional funding, construction ground to a halt, and Tesla was forced to abandon his most ambitious project. Meanwhile, Marconi's company was establishing wireless stations around the world and building a profitable business based on ship-to-shore communication and international telegraphy.

The Forgotten Pioneers

While Tesla and Marconi dominated headlines and captured public attention, other inventors around the world were also making crucial contributions to the development of wireless communication. These forgotten pioneers often lacked the resources, social connections, or publicity skills needed to compete with Tesla and Marconi, but their technical contributions were no less significant.

In Russia, Alexander Popov was conducting wireless experiments that paralleled Marconi's work. Popov demonstrated his wireless receiver at the Russian Physical and Chemical Society in May 1895, several months before Marconi's first successful transmissions. Popov's approach was more scientifically rigorous than Marconi's, and he made important contributions to antenna design and signal detection. However, Popov was primarily interested in scientific research rather than commercial development, and he lacked the business connections needed to transform his discoveries into a practical communication system.

In India, Jagadish Chandra Bose was conducting sophisticated experiments with electromagnetic waves that demonstrated remarkable foresight about the future of wireless technology. Bose developed extremely sensitive detectors that could respond to electromagnetic waves, and he demonstrated wireless transmission over distances of several kilometers. His work on semiconductor detectors anticipated technologies that wouldn't become practical for several decades. Like Popov, however, Bose was more interested in pure scientific research than in commercial applications.

Reginald Fessenden, working in North America, made crucial contributions to the development of continuous wave transmission and voice radio. While Tesla and Marconi were focused on wireless telegraphy using spark transmitters, Fessenden recognized that continuous waves would be necessary for transmitting voice and music. His experiments with alternators and arc transmitters laid the foundation for amplitude modulation (AM) radio broadcasting that would transform popular entertainment in the 1920s.

The Business of Wireless

The contrast between Tesla's visionary approach and Marconi's practical business focus became increasingly apparent as the wireless industry developed. Marconi understood that successful technology required not only technical innovation but also patent protection, manufacturing capability, and market development. He established the Marconi Wireless Telegraph Company in 1897 and systematically built a global network of wireless stations and services.

Marconi's business strategy was based on controlling key patents and establishing a monopoly over wireless communication services. His company not only manufactured wireless equipment but also operated wireless stations and provided communication services to customers. This vertically integrated approach allowed Marconi to capture value at every level of the wireless industry and provided the cash flow needed to fund continued technical development.

Tesla, by contrast, seemed more interested in pushing the boundaries of what was possible than in building a sustainable business around wireless technology. His patents were often too broad or too theoretical to provide effective commercial protection, and his tendency to make grandiose claims about future capabilities sometimes undermined his credibility with potential investors and customers.

The difference in business approaches had profound consequences for both inventors' legacies. Marconi's focused commercial strategy enabled him to build a global wireless empire and establish himself as the practical pioneer of radio communication. Tesla's broader vision, while ultimately more prescient, failed to generate the resources needed to compete effectively with Marconi's more pragmatic approach.

The 1943 Supreme Court Decision

The legal battle between Tesla and Marconi continued long after both inventors had made their major contributions to wireless technology. Marconi died in 1937, having received numerous honors including the 1909 Nobel Prize in Physics (shared with Karl Ferdinand Braun). Tesla died in 1943, still largely unrecognized for his fundamental contributions to wireless communication.

Six months after Tesla's death, the United States Supreme Court issued a decision that dramatically altered the historical record of radio's invention. In Marconi Wireless Telegraph Co. of America v. United States, the Court ruled that several of Marconi's key patents were invalid because they had been anticipated by earlier work, including Tesla's 1900 patent for wireless transmission systems.

The Supreme Court's decision was not primarily about determining who invented radio—as the Court explicitly stated, "Marconi's reputation as the man who first achieved successful radio transmission rests on his original patent... which is not here in question". Rather, the case involved specific patent claims related to four-tuned circuit systems that multiple inventors had developed independently.

Nevertheless, the Court's analysis affirmed that Tesla had indeed developed crucial aspects of wireless communication before Marconi. The decision noted that Tesla's patent number 645,576 had described four tuned circuits before Marconi's competing patent, and that Tesla's work demonstrated "an intimate understanding of the mathematical and physical principles underlying radio communication."

The Supreme Court also recognized the contributions of other inventors, including John Stone Stone and Oliver Lodge, whose patents were also found to anticipate aspects of Marconi's later work. This broader recognition revealed that the invention of radio was not the work of a single genius but rather the result of parallel developments by multiple inventors working on similar problems.

The Historical Verdict

The Tesla-Marconi rivalry illustrates the complex relationship between scientific discovery, technical innovation, and historical recognition. Tesla's theoretical insights and early demonstrations clearly established many of the fundamental principles of wireless communication, but his inability to transform these insights into practical systems limited his immediate impact on the development of radio technology.

Marconi's focused approach to solving specific technical problems, combined with his superior business and marketing skills, enabled him to build the first commercially successful wireless communication systems. His practical achievements, from the transatlantic transmission to the establishment of global wireless services, had immediate and tangible impacts on human communication and transportation.

Yet the 1943 Supreme Court decision suggests that the historical record may have been distorted by Marconi's superior publicity and legal resources. Tesla's fundamental contributions to wireless technology were not adequately recognized during his lifetime, in part because his broader vision made it difficult for him to focus on the specific technical and commercial challenges that Marconi addressed so effectively.

The truth is that both inventors made crucial contributions to the development of radio communication, and the technology that emerged was built on insights from Tesla, Marconi, and numerous other pioneers whose work is often overlooked. Tesla provided much of the theoretical foundation and demonstrated key principles, Marconi solved critical practical problems and built the first successful commercial systems, while others like Popov, Bose, and Fessenden contributed essential technical innovations that made modern radio possible.

The Legacy of Competition

The Tesla-Marconi rivalry had profound implications for the development of wireless technology that extended far beyond the question of who deserves credit as radio's inventor. Their competition spurred rapid technical innovation as each inventor sought to outdo the other's achievements. The patent battles, while contentious, helped establish the legal framework for protecting intellectual property in the emerging wireless industry.

Perhaps most importantly, the rivalry demonstrated two different but complementary approaches to technological innovation. Tesla's visionary approach identified long-term possibilities and established theoretical foundations that would prove valuable for decades. Marconi's practical approach solved immediate problems and created the commercial infrastructure needed to transform laboratory demonstrations into global communication systems.

Modern wireless technology reflects contributions from both approaches. The theoretical insights that Tesla provided about tuned circuits, remote control, and wireless power transmission have become fundamental to countless modern applications. The business and engineering disciplines that Marconi pioneered—systematic patent protection, integrated manufacturing and service provision, and focus on customer needs—remain essential to successful technology companies today.

The great inventor race between Tesla and Marconi was ultimately won not by either individual alone, but by the broader community of inventors, engineers, and entrepreneurs who built upon their discoveries. The invisible waves that Hertz had first demonstrated in his laboratory had been transformed into practical communication systems that would connect the world, thanks to the combined efforts of these pioneering minds. The age of radio had truly begun, emerging from the spark of competition between two extraordinary inventors whose rivalry had pushed the boundaries of what seemed possible.

Chapter 4: From Dots and Dashes to Voices

The crackling sparks that danced across Heinrich Hertz's laboratory apparatus and powered Guglielmo Marconi's early wireless systems carried within them a fundamental limitation that would frustrate inventors and operators for decades. These spark-gap transmitters, marvels of nineteenth-century engineering though they were, could only produce brief, intermittent bursts of electromagnetic energy—perfect for sending the dots and dashes of Morse code, but utterly inadequate for transmitting the continuous waves needed to carry human speech. The wireless world of the 1890s was a world of staccato pulses, of telegraphic brevity, of communication reduced to its most essential elements.

Yet even as Marconi's spark transmitters began connecting ships to shore and continent to continent, visionary inventors were already looking beyond the limitations of wireless telegraphy toward a far more ambitious goal: the wireless transmission of the human voice itself. This transition from dots and dashes to voices would prove to be one of the most technically challenging leaps in the history of communications, requiring fundamental innovations in transmitter design, modulation techniques, and signal detection that would lay the foundation for all modern radio broadcasting.

The Morse Code Monopoly

In the final years of the nineteenth century, wireless communication was synonymous with wireless telegraphy. Marconi's spark-gap transmitters, the most successful of the early wireless systems, operated by creating violent electrical discharges that generated brief pulses of electromagnetic energy. These intermittent bursts of radio waves were ideally suited for transmitting the dots and dashes of Morse code—short pulses for dots, longer pulses for dashes, with silence between characters and words.

The Morse code system, invented by Samuel Morse in the 1830s for wire telegraphy, had proven remarkably effective for reducing human language to a series of simple on-off signals that could be transmitted over even the most primitive communication channels. Each letter of the alphabet was represented by a unique combination of dots and dashes: A was dot-dash, B was dash-dot-dot-dot, C was dash-dot-dash-dot, and so forth. Skilled telegraph operators could send and receive Morse code at remarkable speeds, achieving transmission rates of 25 to 40 words per minute with high accuracy.

But Morse code had significant limitations that became increasingly apparent as wireless technology began to spread beyond its initial military and maritime applications. First, it required specialized training—both senders and receivers needed to memorize the Morse alphabet and develop the manual dexterity to operate telegraph keys efficiently. Second, it was slow compared to normal speech, creating a bottleneck that limited the amount of information that could be transmitted in a given time. Third, it lacked the nuance and emotional content of human speech, reducing all communication to bare factual content.

Most significantly, Morse code created a barrier between wireless technology and the general public. While telegraph operators could communicate efficiently using dots and dashes, ordinary people found the system intimidating and inaccessible. For wireless communication to reach its full potential, inventors would need to find a way to transmit actual human speech—to make radio as natural and intuitive as a telephone conversation.

The Maritime Imperative

The urgency of developing voice transmission was particularly acute in maritime communications, where wireless telegraphy was rapidly becoming an essential safety technology. Ships at sea faced unique communication challenges that made the limitations of Morse code especially problematic. In emergency situations, when every second counted, the delays inherent in encoding and decoding telegraphic messages could mean the difference between life and death.

The maritime industry had been among the first to recognize the value of Marconi's wireless telegraphy systems. Ships equipped with wireless apparatus could maintain contact with shore stations and with other vessels, enabling coordination of shipping schedules, weather reporting, and emergency assistance. By 1900, major shipping lines were beginning to install wireless equipment on their vessels, and coastal stations were being established to provide communication services to maritime traffic.

But the limitations of wireless telegraphy became painfully apparent in emergency situations. When a ship encountered trouble at sea, critical information needed to be conveyed quickly and accurately: the vessel's exact position, the nature of the emergency, the number of people on board, and the urgency of the situation. Transmitting this information in Morse code was time-consuming and error-prone, especially when operators were working under extreme stress.

Furthermore, many ship's officers were not trained telegraph operators. While they could manage routine communications with the assistance of specialized wireless operators, emergency situations often required immediate communication by whoever was available. The ability to transmit voice messages directly, without the need for encoding and decoding, would dramatically improve the effectiveness of maritime emergency communications.

The commercial shipping industry also saw clear benefits in voice transmission. Radio telephone conversations between ship and shore could handle complex logistical arrangements, weather discussions, and business negotiations far more efficiently than the abbreviated style required by telegraphic communication. Voice transmission would make wireless communication more accessible to ordinary mariners and would expand the range of applications for shipboard radio systems.

Fessenden's Revolutionary Vision

The technical challenge of transmitting voice by radio was immediately apparent to anyone who understood the physics of electromagnetic waves. Voice transmission required continuous waves that could be modulated—varied in amplitude or frequency—to carry the complex waveforms of human speech. The spark-gap transmitters that dominated early wireless systems produced only brief bursts of radio energy, making voice modulation impossible.

Canadian-American inventor Reginald Fessenden recognized this fundamental limitation earlier than most of his contemporaries.

Working for the United States Weather Bureau in the late 1890s, Fessenden came to the conclusion that wireless telegraphy represented only the first primitive step toward true wireless communication. He envisioned a wireless system that could transmit not only voice but also music, enabling radio to become a medium for entertainment and education as well as point-to-point communication.

Fessenden's approach to the voice transmission problem was characteristically systematic and technically sophisticated. He understood that successful voice transmission would require three fundamental innovations: a continuous-wave transmitter that could produce steady radio signals rather than intermittent sparks, a modulation system that could impose voice signals onto the radio carrier wave, and sensitive receivers that could detect and demodulate the transmitted signals to recover the original voice content.

The first challenge—developing a continuous-wave transmitter—proved to be the most technically demanding. Fessenden began experimenting with high-speed alternators, essentially AC generators that could produce alternating current at radio frequencies. Unlike spark-gap transmitters, which created brief pulses of electromagnetic energy, these alternators could generate "pure sine waves" and produce "a continuous train of radiant waves of substantially uniform strength"—exactly what was needed for voice transmission.

Working on Cobb Island, Maryland, Fessenden dedicated himself to perfecting this continuous-wave technology. The technical challenges were formidable: radio-frequency alternators needed to operate at much higher speeds than conventional power generators, requiring precision engineering and exotic materials. The alternators also needed to be extremely stable, as any variations in frequency or amplitude would distort the transmitted voice signals.

The First Words Through the Ether

By the fall of 1900, Fessenden had achieved a breakthrough that would forever change the nature of wireless communication. Using his continuous-wave transmitter system, he successfully transmitted speech over a distance of about 1.6 kilometers (one mile)—the first successful audio transmission using radio signals in history. The technical achievement was remarkable, but the practical quality left much to be desired. According to Fessensen's own accounts, "the sound transmitted was far too distorted to be commercially practical."

Despite the poor audio quality, Fessenden's initial success proved that voice transmission by radio was possible. The distortion problems that plagued these early experiments were primarily due to limitations in the modulation and detection systems rather than fundamental flaws in the continuous-wave approach. With continued development, Fessenden was confident that these technical problems could be solved.

The significance of this first voice transmission extended far beyond its immediate practical applications. Fessenden had demonstrated that radio waves could carry information far more complex than the simple on-off signals of wireless telegraphy. If human speech could be transmitted, then music, too, should be possible. The implications for entertainment, education, and mass communication were staggering, though few people at the time grasped the full potential of Fessenden's achievement.

Fessenden's success also validated his theoretical understanding of radio wave propagation and modulation. His continuous-wave approach represented a fundamental departure from the spark-gap methods that dominated contemporary wireless systems. While other inventors were focused on improving spark transmitters and developing more sensitive coherer receivers, Fessenden had recognized that voice transmission required entirely different technical approaches.

The Christmas Eve Miracle

The culmination of Fessenden's work on voice transmission came on Christmas Eve, 1906, in one of the most famous experiments in radio history. From his transmitting station at Brant Rock, Massachusetts, Fessenden broadcast what is widely considered the first radio program ever transmitted. Using an Alexanderson alternator and rotary spark-gap transmitter, he sent out voice and music to ships at sea throughout the Atlantic Ocean.

The Christmas Eve broadcast was carefully planned to demonstrate the full potential of voice radio transmission. Fessenden began by playing "O Holy Night" on his violin, the haunting melody carrying across hundreds of miles of ocean to astonished radio operators on ships who had never heard music transmitted wirelessly. He followed this musical performance by reading a passage from the Bible, his voice reaching listeners who were scattered across the Atlantic shipping lanes.

The impact of this broadcast on its listeners was profound and immediate. Ship radio operators, accustomed to receiving only the staccato dots and dashes of Morse code, were amazed to hear actual human speech and music emerging from their wireless receivers. Many initially thought their equipment was malfunctioning or that they were somehow receiving signals from a nearby ship. The concept of broadcast transmission—one transmitter sending the same signal to multiple receivers simultaneously—was so novel that many listeners couldn't immediately comprehend what they were experiencing.

From a technical standpoint, the Christmas Eve broadcast represented a remarkable achievement in radio engineering. Fessenden's transmitter had to generate sufficient power to reach ships hundreds of miles at sea, while maintaining the signal quality necessary for intelligible voice and music reproduction. The modulation system had to impose the complex waveforms of speech and music onto the radio carrier wave without excessive distortion, and the transmission had to be stable enough to maintain quality throughout the entire broadcast.

The success of the Christmas Eve broadcast validated Fessenden's vision of radio as a medium for entertainment and mass communication. Unlike wireless telegraphy, which was primarily a point-to-point communication technology serving specialized users, voice radio had the potential to reach large audiences simultaneously. This broadcast capability would prove to be crucial to radio's evolution from a specialized communication tool to a mass medium that would transform popular culture.

De Forest and the Audion Revolution

While Fessenden was pioneering continuous-wave voice transmission, another American inventor was approaching the same problem from a different angle. Lee de Forest, working in New York City, was experimenting with arc transmitters that could also produce continuous waves suitable for voice modulation. De Forest's approach differed from Fessenden's alternator-based system, but it addressed the same fundamental limitation of spark-gap transmitters.

De Forest's arc transmitter created "a steady continuous wave signal that could be used for amplitude modulated (AM) audio transmissions". Unlike the intermittent pulses produced by spark transmitters, arc transmitters maintained a constant radio signal that could be modulated to carry voice and music. In February 1907, de Forest demonstrated this capability by transmitting electronic telharmonium music from his laboratory station in New York City, followed later that year by tests that included vocalist Eugenia Farrar singing "I Love You Truly."

Perhaps more significantly for the future of radio, de Forest had invented the audion tube in 1906—a three-electrode vacuum tube that could amplify weak radio signals. This invention would prove to be crucial to the development of practical radio receivers, though its full potential wouldn't be realized until the 1920s. The audion's ability to amplify signals meant that voice radio transmissions could be received at much greater distances and with much better quality than was possible with the primitive detectors available in the early 1900s.

De Forest's work in July 1907 included ship-to-shore voice transmissions during the Annual Inter-Lakes Yachting Association Regatta on Lake Erie. These tests demonstrated that voice radio could be used for practical maritime communications, transmitting race reports from the steam yacht Thelma to shore stations. This application showed that voice radio wasn't just a laboratory curiosity but could serve real communication needs in maritime environments.

The parallel development of voice transmission systems by both Fessenden and de Forest illustrated the technical readiness of the radio art for this crucial advance. Multiple inventors working independently had recognized the limitations of spark-gap telegraphy and had developed different but effective solutions to the voice transmission problem. This convergent evolution suggested that voice radio represented a natural and inevitable step in the development of wireless technology.

Technical Challenges and Solutions

The transition from wireless telegraphy to voice transmission required solving several fundamental technical problems that had not been apparent in the earlier dot-and-dash era. The first challenge was generating continuous waves with sufficient stability and purity to carry voice modulation without distortion. Early continuous-wave generators, whether alternators or arc transmitters, were prone to frequency drift and amplitude variations that could severely degrade voice quality.

Fessenden's solution involved developing high-frequency alternators that could operate at radio frequencies while maintaining excellent stability. These machines represented remarkable achievements in precision engineering, requiring careful attention to bearing design, magnetic circuit optimization, and mechanical balance. The alternators had to run at extremely high speeds—often tens of thousands of revolutions per minute—while maintaining frequency stability measured in fractions of a percent.

The second major challenge was developing effective modulation techniques that could impose voice signals onto radio carrier waves without excessive distortion. Early amplitude modulation systems were relatively crude, often producing significant harmonic distortion that made speech difficult to understand and music unpleasant to hear. Improving modulation quality required careful attention to circuit design, component selection, and operating parameters.

Detection and demodulation at the receiving end presented equally challenging problems. The sensitive coherers used for wireless telegraphy were not suitable for voice reception, as they could only detect the presence or absence of radio signals, not their amplitude variations. Voice radio required detectors that could follow the amplitude changes of modulated signals and convert them back into audio frequencies that could drive headphones or loudspeakers.

Early voice radio systems used crystal detectors—pieces of mineral crystals like galena or pyrite that could rectify radio-frequency signals and extract the audio modulation. These detectors were finicky and required constant adjustment, but they were sensitive enough to enable voice reception over significant distances. The development of more stable and sensitive detection systems would remain an active area of research for decades.

The Titanic: A Catalyst for Change

The most dramatic demonstration of wireless communication's life-saving potential came during the tragic sinking of the RMS Titanic on the night of April 14-15, 1912. The disaster, which claimed over 1,500 lives, also highlighted both the capabilities and limitations of wireless telegraphy in emergency situations. The Titanic's wireless operators, Jack Phillips and Harold Bride, worked heroically to transmit distress calls using Morse code, ultimately enabling the rescue of over 700 survivors by the steamship Carpathia.

The Titanic disaster galvanized public and government attention on maritime wireless communication in ways that peacetime demonstrations never could. The stark reality that wireless distress calls had meant the difference between life and death for hundreds of people made it clear that radio was not merely a convenience but an essential safety technology. In the aftermath of the disaster, maritime wireless regulations were strengthened, and the installation of wireless equipment on passenger vessels became mandatory.

However, the Titanic disaster also revealed the limitations of wireless telegraphy in emergency situations. The transmission of detailed information about the ship's condition, the number of people on board, and the urgency of the situation required lengthy Morse code messages that consumed precious time. Voice transmission would have enabled much more efficient emergency communication, allowing complex information to be conveyed quickly and accurately.

The disaster also highlighted the importance of continuous radio watch-keeping. Several ships in the Titanic's vicinity had wireless equipment but were not maintaining 24-hour radio watches, meaning that distress calls went unheard. The tragedy led to international agreements requiring passenger ships to maintain continuous wireless watches and to carry backup power systems for radio equipment.

Perhaps most importantly, the Titanic disaster demonstrated the value of wireless communication to the general public in terms that everyone could understand. The dramatic rescue operation, coordinated entirely by wireless messages, showed that radio was not just an interesting technical novelty but a technology that could save lives. This public awareness would prove crucial to radio's rapid adoption in the years following the disaster.

The Dawn of Broadcasting

The technical breakthroughs achieved by Fessenden, de Forest, and other voice radio pioneers established the foundation for what would become radio broadcasting, though the concept of mass audience radio was still years in the future. The early voice transmission experiments were primarily focused on point-to-point communication—replacing or supplementing wireless telegraphy for specific applications like maritime communications or military use.

However, the broadcast nature of radio transmission was inherent in the technology from the beginning. Unlike telephone systems, which established dedicated circuits between individual users, radio transmission was inherently a one-to-many medium. Any receiver within range of a transmitter could potentially receive the transmitted signals, making it possible for one transmitter to serve multiple receivers simultaneously.

Fessenden himself seemed to recognize this broadcast potential, as evidenced by his Christmas Eve 1906 transmission. By sending out music and speech intended for multiple ships simultaneously, he was essentially conducting the world's first radio broadcast. The concept was so novel that there was no established terminology for it—words like "broadcast," "program," and "audience" would not be applied to radio until years later.

The transition from wireless telegraphy to voice transmission also required changes in how radio services were organized and regulated. Wireless telegraphy had been primarily a commercial service, with radio operators sending and receiving messages for paying customers. Voice radio opened up possibilities for entertainment, education, and mass communication that didn't fit the existing business models or regulatory frameworks.

By 1910, the technical elements necessary for practical voice radio were largely in place, though they would not be combined into commercially successful broadcasting systems until after World War I. Continuous-wave transmitters, amplitude modulation techniques, and sensitive receivers had all been demonstrated and were gradually being improved through continued research and development.

The war years brought intensive development of radio technology for military applications, leading to significant improvements in transmitter power, receiver sensitivity, and signal quality. Military requirements for reliable long-distance voice communication spurred innovations that would later benefit commercial radio development. The vacuum tube technology pioneered by de Forest and others was refined and mass-produced for military radio systems, laying the groundwork for the vacuum tube receivers that would dominate radio in the 1920s.

During the mid-1920s, amplifying vacuum tubes revolutionized both radio receivers and transmitters. The development of more sophisticated vacuum tube circuits enabled receivers with much better sensitivity and selectivity, making it possible to receive distant stations clearly even in the presence of local interference. Transmitter improvements allowed voice signals to be broadcast over much greater distances with higher quality than the early experimental systems had achieved.

The invention of the superheterodyne receiver principle provided another crucial advance in radio technology. This receiver design, which became standard for radio equipment, offered superior selectivity and sensitivity compared to earlier receiver types. The superheterodyne principle enabled listeners to tune in weak distant stations while rejecting interference from nearby transmitters, making practical radio broadcasting possible in areas with multiple competing stations.

The Social Transformation Begins

The transition from dots and dashes to voices represented more than just a technical evolution—it marked the beginning of radio's transformation from a specialized communication tool into a mass medium that would reshape society. Voice radio made wireless communication accessible to ordinary people who had neither the time nor inclination to learn Morse code. This accessibility was crucial to radio's eventual role in entertainment, news, education, and mass communication.

The implications of this transition extended far beyond the immediate practical benefits of voice communication. For the first time in human history, it became possible for a single person to speak simultaneously to audiences scattered across vast geographic areas. This capability would prove to have profound effects on politics, culture, commerce, and social organization.

Voice radio also began to break down the isolation that had historically characterized rural and remote areas. Farmers, sailors, and residents of distant settlements could now receive news, weather information, and entertainment from urban centers. This connectivity would gradually erode the sharp distinctions between urban and rural cultures that had persisted throughout human history.

The technical foundation laid by Fessenden, de Forest, and their contemporaries made possible the radio broadcasting boom that would sweep the world in the 1920s. Their work on continuous-wave transmission, amplitude modulation, and sensitive detection established the basic principles that would guide radio development for decades to come. The transformation from dots and dashes to voices was complete; the age of radio broadcasting was about to begin.

The wireless waves that had first carried only the stark simplicity of Morse code now danced with the full complexity of human expression—laughter and tears, music and poetry, news and entertainment. The invisible medium that connected the world had found its voice, and that voice would soon reshape human civilization in ways that even the most visionary pioneers could scarcely have imagined. The foundation years of radio were drawing to a close, and the broadcasting revolution was about to begin.

Part II: The Broadcasting Revolution (1920-1950)

Chapter 5: Radio Comes Home - The Birth of Broadcasting

The year 1920 dawned on a world forever changed by the Great War, where millions had witnessed the devastating power of modern technology turned toward destruction. Yet in the quiet laboratories and amateur workshops scattered across America, a different kind of revolution was brewing—one that would harness the invisible waves of the electromagnetic spectrum not for warfare, but for bringing families together around a simple wooden box that could capture voices and music from the very air itself. The transition from experimental wireless communication to regular radio broadcasting represented one of the most dramatic technological and social transformations in human history, turning radio from a specialized tool used by trained operators into an intimate companion that would reshape domestic life, popular culture, and the very fabric of society.

The journey from Fessenden's pioneering Christmas Eve broadcast of 1906 to the establishment of regular broadcasting services was neither straight nor swift. For more than a decade after Fessenden demonstrated that voices could travel through space, radio remained primarily the domain of experimenters, maritime operators, and military communications specialists. The general public remained largely unaware of radio's potential as an entertainment medium, viewing it as a mysterious technical novelty rather than a practical technology that might enter their daily lives. But beneath this apparent stagnation, crucial developments were taking place that would soon make radio broadcasting not just possible but inevitable.

The Seeds of Broadcasting: Early Experimental Stations

The conceptual foundation for radio broadcasting emerged gradually from the experimental work of dedicated pioneers who recognized that radio's one-to-many transmission capability could serve purposes far beyond point-to-point communication. Unlike telephones, which connected individual callers, or telegraphs, which carried messages between specific points, radio transmission was inherently a broadcast medium—one transmitter could simultaneously reach any number of receivers within its coverage area.

The University of Wisconsin's experimental station 9XM, which began regular transmissions in 1916, is often credited as one of the first stations to establish anything resembling regular broadcasting. Operating from the university's physics department, 9XM initially focused on transmitting weather reports and market information to farmers throughout Wisconsin. The station's call letters would later become WHA, and it continues broadcasting today as one of America's oldest radio stations, providing a direct link to the earliest days of organized broadcasting.

These early experimental broadcasts served practical rather than entertainment purposes. Weather information was crucial for agricultural communities, while market reports helped farmers make informed decisions about when and where to sell their crops. The broadcasts were typically brief and factual, consisting of telegraphic-style information read aloud rather than the conversational programming that would characterize later broadcasting. Most receivers were still primitive crystal sets operated by technically minded amateurs who could tolerate the inconveniences of finicky equipment and unpredictable reception quality.

The format of these early broadcasts reflected their experimental nature and limited audience. Transmissions were typically scheduled for specific times and announced in advance through technical publications and amateur radio networks. The content was sparse and utilitarian: weather conditions, temperature readings, barometric pressure, and occasionally news of particular interest to the rural communities that made up most of the listening audience. Musical programming, when it occurred, was usually limited to gramophone records or live performances by station personnel who happened to have musical ability.

Despite their primitive nature, these early broadcasts demonstrated radio's potential to serve community needs in ways that no previous communication technology could match. A single weather broadcast could simultaneously inform hundreds or thousands of farmers across a broad geographic area, providing timely information that could influence crucial agricultural decisions. Market reports could help rural producers get better prices for their crops by informing them about supply and demand conditions in distant cities.

The Amateur Radio Renaissance

The period between 1910 and 1920 witnessed an extraordinary growth in amateur radio activity that created the technical foundation and operator base necessary for broadcasting to succeed. Thousands of young men (and a few women) across America and Europe built their own radio equipment, experimented with transmission and reception, and created informal networks of communication that spanned continents. These radio amateurs, or "hams" as they came to be known, developed much of the practical knowledge about antenna design, circuit construction, and operating procedures that would prove essential when commercial broadcasting began.

Amateur radio operators were driven by a combination of technical curiosity and social connection that made radio more than just a hobby—it became a passion that consumed their spare time and disposable income. Young men would spend hours in basement workshops winding coils, adjusting capacitors, and testing circuits, driven by the dream of communicating with distant stations. The thrill of receiving a faint signal from hundreds or thousands of miles away, perhaps from another amateur operator struggling with similar equipment challenges, created a sense of community and shared purpose that transcended geographic boundaries.

The amateur radio community developed its own culture, complete with specialized terminology, operating procedures, and ethical codes that emphasized experimentation, mutual assistance, and technical excellence. Amateur operators shared information freely, publishing circuit diagrams and construction details in magazines like "QST" and "Modern Electrics." This open exchange of technical knowledge accelerated radio development and created a pool of experienced operators who would later staff the commercial broadcasting stations that emerged in the 1920s.

World War I temporarily disrupted amateur radio activity as governments seized control of radio facilities for military use, but the war also spurred rapid technical development that would benefit civilian radio after the conflict ended. Military requirements for reliable long-distance communication led to improvements in transmitter power, receiver sensitivity, and circuit design. When the war ended and radio restrictions were lifted, amateur operators returned to their hobby with access to surplus military equipment and enhanced technical knowledge gained through wartime service.

The postwar amateur radio boom created exactly the conditions necessary for broadcasting to flourish. Thousands of experienced operators possessed the technical skills to build and maintain radio equipment, while thousands more were interested in radio but lacked the time or expertise to become active amateurs. This second group represented the potential audience for broadcasting—people who wanted to enjoy radio's capabilities without having to master its technical complexities.

Crystal Sets and the Democratization of Radio Reception

The key technology that made radio broadcasting practical for ordinary consumers was the crystal detector, a simple device that could convert radio frequency signals into audio frequencies without requiring batteries, vacuum tubes, or other complex components. Crystal sets, as these receivers came to be known, consisted of little more than a coil of wire (the inductor), a variable capacitor for tuning, a crystal detector for demodulation, and headphones for listening. The entire receiver could be built for a few dollars using readily available materials, bringing radio reception within reach of middle-class families for the first time.

The crystal detector itself was a marvel of simplicity that worked through principles not fully understood at the time. A small piece of galena, pyrite, or other mineral crystal was mounted in a holder and contacted by a thin wire called a "cat's whisker." When radio waves induced alternating currents in the receiver's antenna circuit, the crystal-wire contact acted as a primitive semiconductor diode, allowing current to flow in only one direction and thereby converting the amplitude-modulated radio signal into varying direct current that could drive headphones.

Crystal sets had significant limitations that seem almost quaint by modern standards. They produced no amplification, meaning that only relatively strong local signals could be received clearly. The crystal detector had to be adjusted frequently by moving the cat's whisker to find a sensitive spot on the crystal surface, a process that required patience and practice. Reception quality varied dramatically with weather conditions, time of day, and seasonal atmospheric changes. Despite these limitations, crystal sets made radio accessible to people with no technical training and created the mass audience that broadcasting required.

The construction of crystal sets became a popular hobby that introduced thousands of people to basic radio principles. Magazines and newspapers published construction articles with detailed instructions for building simple receivers, while small businesses sprang up to supply coils, crystals, headphones, and other components. Father-and-son crystal set projects became a common weekend activity, creating family bonds around radio technology and fostering technical interest among young people who would later pursue careers in electronics and broadcasting.

Crystal set performance could be improved through careful attention to construction details and operating technique. Longer antennas collected more signal energy, while good ground connections reduced noise and improved sensitivity. Multiple-circuit receivers with several tuned stages could provide better selectivity, allowing listeners to separate signals from different stations. Experienced crystal set operators learned to recognize the characteristics of different stations and could often identify transmitters by their distinctive audio qualities or transmission patterns.

The social impact of crystal sets extended far beyond their technical capabilities. For the first time, ordinary families could receive entertainment, news, and information from distant cities without leaving their homes. Rural families, in particular, found that radio connected them to urban culture and current events in ways that had never before been possible. The shared experience of listening to the same programs created common cultural references and helped break down the isolation that had historically characterized rural life.

The Westinghouse Breakthrough: KDKA and Commercial Broadcasting

The transformation of radio from experimental novelty to mass medium occurred with stunning rapidity between 1920 and 1922, triggered by the Westinghouse Electric Corporation's decision to establish regular broadcasting as a means of creating demand for radio receivers. The company's executives recognized that radio equipment sales would remain limited as long as there was little worth listening to, creating a classic chicken-and-egg problem that could only be solved by providing both equipment and programming simultaneously.

Westinghouse's solution was elegantly simple: establish a broadcasting station that would provide regular programming to stimulate receiver sales, then manufacture and market radio receivers to capture the demand that broadcasting created. The company applied for and received a commercial broadcasting license, establishing station KDKA in Pittsburgh, Pennsylvania, which began regular operations on November 2, 1920, with live coverage of the Harding-Cox presidential election returns.

The choice of election coverage for KDKA's inaugural broadcast was strategically brilliant, combining news value with the kind of immediate, time-sensitive information that radio could provide better than any other medium. While newspaper readers would have to wait for morning editions to learn election results, radio listeners could follow the vote count as it progressed through the evening. The broadcast demonstrated radio's unique capability for providing real-time information and established news as one of broadcasting's fundamental programming categories.

KDKA's success exceeded Westinghouse's most optimistic expectations. Within weeks of the station's debut, hundreds of letters arrived from listeners scattered across a broad geographic area, proving that regular broadcasting could attract and maintain a substantial audience. The publicity generated by KDKA's broadcasts created enormous demand for radio receivers, validating Westinghouse's business model and inspiring other companies to establish their own broadcasting stations.

The technical quality of KDKA's broadcasts was remarkably good for the era, reflecting Westinghouse's engineering expertise and substantial financial investment. The station used a 100-watt transmitter that provided reliable coverage throughout the Pittsburgh metropolitan area and could be received at much greater distances under favorable conditions. The studio facilities, while primitive by later standards, were acoustically treated to minimize echo and background noise, while the programming included live music, news, weather reports, and talks on subjects of general interest.

KDKA's programming philosophy emphasized variety and broad appeal rather than the specialized technical content that had characterized earlier experimental broadcasts. The station featured live performances by local musicians, talks by university professors and community leaders, play-by-play coverage of sporting events, and regular news bulletins that kept listeners informed about current events. This diverse programming attracted listeners from all segments of society and demonstrated radio's potential as a mass entertainment medium.

The Broadcasting Explosion of 1921-1922

The success of KDKA triggered an extraordinary explosion of broadcasting activity that transformed the American media landscape almost overnight. Companies across the country rushed to establish their own broadcasting stations, recognizing the promotional value of radio programming and the potential profits from receiver sales. By the end of 1922, over 500 broadcasting stations were operating in the United States, creating a chaotic but vibrant radio environment that offered listeners an unprecedented variety of programming choices.

This rapid expansion was facilitated by the relatively simple technical requirements for establishing a broadcasting station. Unlike modern broadcasting, which requires sophisticated equipment and substantial financial investment, early radio stations could be established with relatively modest resources. A few hundred dollars could purchase a low-power transmitter, while studio facilities might consist of little more than a microphone in a converted office or store room. Many stations were operated as promotional ventures by radio dealers, newspapers, or department stores rather than as independent businesses expected to generate revenue through advertising.

The proliferation of broadcasting stations created unprecedented opportunities for musicians, speakers, and entertainers who had previously been limited to live audiences in specific geographic locations. Local talent could now reach listeners across entire metropolitan areas, while particularly popular performers might be heard over networks of stations covering multiple cities. Radio began attracting professional entertainers from vaudeville, theater, and the recording industry, gradually raising the quality and sophistication of broadcast programming.

However, the rapid growth of broadcasting also created serious technical and regulatory problems. With hundreds of stations operating on a limited number of available frequencies, interference became a persistent problem that degraded reception quality and frustrated listeners. The federal government's regulatory framework was inadequate for managing the complex frequency coordination required by broadcasting, leading to conflicts between stations and calls for more sophisticated regulatory approaches.

The programming diversity of the early 1920s was remarkable, reflecting the experimental nature of broadcasting and the absence of established formats or commercial constraints. Stations broadcast everything from classical music concerts to farm market reports, from children's stories to university lectures. Some stations specialized in particular types of content—religious programming, foreign language broadcasts, or educational material—while others attempted to serve general audiences with varied programming throughout the day.

Radio Enters the American Home

The period from 1921 to 1925 witnessed radio's transformation from a technical curiosity operated by trained amateurs to a domestic appliance that became central to American family life. This transition required not only improvements in receiver technology and programming quality but also fundamental changes in how families organized their leisure time and domestic space. Radio reception evolved from an individual activity requiring headphones and technical skill to a shared family experience centered around loudspeaker-equipped receivers that anyone could operate.

The development of practical vacuum tube receivers was crucial to radio's domestication. Unlike crystal sets, which provided no amplification and required headphones for listening, vacuum tube receivers could drive loudspeakers and provide sufficient volume for family listening. Early vacuum tube receivers were expensive and required regular maintenance, including replacement of the vacuum tubes themselves, but they offered much better performance than crystal sets and could receive distant stations reliably.

The introduction of manufactured radio receivers marked another crucial step in broadcasting's development. While early radio enthusiasts had built their own equipment, the mass market for broadcasting required reliable, easy-to-operate receivers that could be purchased and used by people with no technical training. Companies like Atwater Kent, Philco, and RCA began manufacturing complete radio receivers that combined all necessary components in attractive wooden cabinets designed to complement home furnishings rather than look like laboratory equipment.

Radio furniture became an important design consideration as receivers grew larger and more elaborate. Early radio sets were typically housed in plain rectangular cabinets that emphasized function over appearance, but as radio became more popular, manufacturers began offering receivers in furniture-style cabinets that could serve as decorative elements in living rooms and family rooms. Console radios became popular, combining radio receivers with phonographs in large floor-standing cabinets that served as the centerpiece of family entertainment.

The social impact of radio on family life was profound and immediate. For the first time, families could enjoy professional entertainment in their own homes without the expense and inconvenience of attending theaters or concerts. Radio programming provided shared experiences that brought family members together around the receiver for evening entertainment, creating new traditions and social patterns that would persist for decades.

Radio also began to influence the rhythm of daily life as families adjusted their schedules to accommodate favorite programs. Dinner times might be shifted to avoid missing popular shows, while household chores could be timed to coincide with morning music programs or afternoon serials. The radio schedule became a organizing principle for domestic life, much as television programming would later influence family routines.

The Economics of Early Broadcasting

The rapid growth of broadcasting in the early 1920s occurred without a clear business model for sustaining programming costs and station operations. Most early broadcasters operated their stations as promotional ventures designed to increase sales of radio equipment, generate publicity for other businesses, or provide community service rather than as profit-making enterprises in their own right. This arrangement worked adequately when broadcasting was limited and costs were modest, but the expansion of programming schedules and improvement of technical quality required more substantial and reliable funding sources.

Radio dealers were among the first commercial broadcasters, using their stations to demonstrate radio receivers and create demand for equipment sales. Department stores operated stations to attract customers and generate publicity, while newspapers established broadcasting stations to complement their print operations and explore the potential of radio journalism. Hotels, theaters, and other entertainment venues used radio broadcasts to promote their primary businesses and reach potential customers beyond their immediate geographic areas.

The question of how to fund broadcasting operations became increasingly urgent as stations expanded their programming and improved their facilities. Volunteer performers and donated programming could sustain only limited broadcast schedules, while the employment of professional talent and the construction of proper studios required substantial ongoing expenses. Various funding models were proposed and tested, including listener subscriptions, municipal support, and endowments from wealthy patrons, but none proved adequate for supporting the level of programming that audiences were beginning to expect.

The introduction of radio advertising, while controversial at first, provided the economic foundation that made commercial broadcasting viable. AT&T's station WEAF in New York City conducted the first experiments with "toll broadcasting" in 1922, selling airtime to advertisers much as newspapers sold space for printed advertisements. Early radio advertising was tasteful and informative by later standards, often consisting of brief announcements about products or services rather than the elaborate commercial productions that would characterize radio in later decades.

The development of radio networks provided economies of scale that made higher-quality programming economically feasible. By linking multiple stations through telephone lines, networks could distribute the same programs to listeners in different cities, spreading the costs of talent and production across larger audiences. The National Broadcasting Company (NBC), established in 1926, was the first permanent radio network, followed by the Columbia Broadcasting System (CBS) and other competitors that created the framework for American broadcasting that persists today.

Programming Innovation and Format Development

The programming experimentation of the early broadcasting era established many of the formats and conventions that would define radio entertainment for decades. With no established precedents to follow, broadcasters and performers had to invent the techniques and formats appropriate for the radio medium, discovering through trial and error what worked effectively on the air and what audiences found appealing.

Music programming dominated early broadcasting, partly because it was relatively inexpensive to produce and partly because it translated well to the radio medium. Live performances by local musicians, ranging from solo pianists to full orchestras, provided hours of daily programming while also showcasing local talent. Dance music became particularly popular as radio listeners discovered they could enjoy the latest songs from popular bands and orchestras in their own homes.

The development of radio drama represented one of broadcasting's most significant creative achievements. Unlike theater, which depended on visual spectacle and stage craft, radio drama had to create entire worlds using only sound—dialogue, music, and sound effects. This limitation forced writers and performers to develop new techniques for storytelling that emphasized character development, atmospheric description, and imaginative sound design. Radio drama would become one of the medium's most distinctive and popular programming forms, producing classics that are still remembered decades later.

News and information programming established radio's role as a medium for current events and public affairs. Radio's ability to provide immediate coverage of breaking news gave it significant advantages over newspapers and magazines, which required hours or days to report developing stories. Sports broadcasting became particularly popular, allowing listeners to follow games and matches as they happened rather than waiting for newspaper accounts the following day.

Educational programming reflected the early idealism about radio's potential to inform and uplift audiences. Universities, museums, and cultural organizations used broadcasting to extend their educational missions beyond their immediate geographic areas. Language lessons, lecture series, and cultural programs attempted to bring higher education into American homes, though these programs generally attracted smaller audiences than entertainment offerings.

The Social Revolution of Shared Experience

Radio's transformation of American social life extended far beyond its role as an entertainment medium. For the first time in history, millions of people could simultaneously experience the same programs, creating shared cultural references and common experiences that transcended geographic, economic, and social boundaries. This mass communication capability had profound implications for American democracy, commercial culture, and social cohesion that are still being felt today.

The democratization of entertainment was among radio's most immediate social effects. Professional-quality music, drama, and comedy that had previously been available only to urban audiences willing to pay theater prices suddenly became accessible to anyone with a radio receiver. Rural families, in particular, found their cultural isolation dramatically reduced as radio brought them the same entertainment that city dwellers enjoyed. This cultural leveling helped break down some of the sharp distinctions between urban and rural life that had characterized American society since colonial times.

Radio also created new forms of celebrity and fame that differed fundamentally from earlier entertainment traditions. Radio performers could become familiar to millions of listeners without ever being seen, creating intimate relationships between entertainers and audiences that were unprecedented in scale and character. The human voice became a vehicle for personality and charisma in ways that had never before been possible, launching careers and creating stars whose appeal was based entirely on their radio presence.

The commercial implications of radio's mass audience were equally profound. National advertising became practical for the first time, as manufacturers could reach consumers across the entire country through network radio programs. This capability accelerated the development of national brands and standardized products, contributing to the homogenization of American consumer culture. Radio advertising also provided a new medium for political communication, enabling candidates and causes to reach voters directly rather than relying entirely on newspapers and public meetings.

Radio's role in creating shared national experiences became particularly apparent during major news events, when millions of listeners could follow developing stories simultaneously. Natural disasters, political events, and international crises became shared national experiences rather than local or regional concerns, contributing to the development of a more unified national consciousness. This capability would prove particularly important during the economic crisis of the 1930s and the global conflict of the 1940s, when radio served as a crucial medium for national unity and morale.

The Technical Infrastructure of Broadcasting

The transformation of radio from experimental communication to mass broadcasting required the development of an entirely new technical infrastructure that could support regular, high-quality programming for large audiences. This infrastructure encompassed not only broadcasting stations and receivers but also the telephone networks, recording technologies, and measurement systems that made professional broadcasting possible.

Broadcasting stations evolved rapidly from the crude experimental facilities of the early 1920s to sophisticated operations that rivaled the technical quality of the best telephone and telegraph systems. Transmitter power increased from the few watts used by early experimental stations to the kilowatts required for reliable regional coverage, while antenna systems became increasingly sophisticated to maximize coverage area and minimize interference with other stations.

Studio design emerged as a specialized field that combined acoustical engineering with architectural planning to create spaces optimized for radio production. Early studios were often converted offices or store rooms with minimal acoustic treatment, but professional broadcasting required carefully designed spaces that controlled reverberation, eliminated unwanted noise, and provided appropriate acoustic environments for different types of programming. The development of specialized microphones, mixing equipment, and recording devices further enhanced the technical quality of broadcast programming.

The telephone network played a crucial role in broadcasting development, providing the communication links necessary for network operations and remote programming. American Telephone and Telegraph Company's (AT&T) high-quality telephone lines enabled the distribution of programs from central studios to affiliated stations across the country, making network broadcasting economically feasible. Remote broadcasts from concert halls, sports venues, and other locations also depended on telephone connections that could carry high-quality audio over long distances.

Recording technology became increasingly important as broadcasting schedules expanded and live programming became more complex. While early broadcasting relied almost entirely on live performances, the development of improved recording and playback equipment made it possible to present recorded programming that maintained professional standards. Transcription services began providing recorded programs to stations that lacked the resources for extensive live programming, democratizing access to high-quality content and expanding programming options for smaller stations.

The Dawn of Mass Media

By the mid-1920s, radio had completed its transformation from experimental curiosity to mass medium, fundamentally altering the landscape of American communications and entertainment. The technology that had begun as a specialized tool for maritime communications and amateur experimentation had become a domestic appliance as common and essential as the telephone or electric lighting. Radio had not only come home—it had become home, the central gathering place around which family life was increasingly organized.

The implications of this transformation extended far beyond entertainment and convenience. Radio had created the first truly mass medium in human history, capable of reaching millions of people simultaneously with the same message. This capability would prove to have profound consequences for politics, commerce, culture, and society that are still unfolding today. The invisible waves that Maxwell had predicted and Hertz had demonstrated had finally found their highest calling: connecting humanity in an invisible web of shared experience that transcended the physical boundaries that had historically separated communities, regions, and nations.

The broadcasting revolution of the 1920s established the foundation for all subsequent developments in electronic media. Television would later build upon radio's technical and business models, while the internet would extend broadcasting's one-to-many communication paradigm into interactive, global networks. But it was radio that first demonstrated the power of electromagnetic waves to create mass audiences, shared experiences, and instantaneous communication across vast distances.

As the 1920s drew to a close, radio had become an established part of American life, present in millions of homes and playing an increasingly important role in politics, commerce, and culture. The crystal sets and headphones of the early broadcasting era were giving way to sophisticated console receivers with powerful loudspeakers, while programming was becoming more professional and diverse. Radio had indeed come home, and in doing so, it had begun to transform the very nature of home itself.

Chapter 6: The Technical Leap - Vacuum Tubes Change Everything

The crackling static that emerged from crystal set headphones in the early 1920s carried with it not just the faint voices and music of distant broadcasting stations, but also the promise of a technological revolution that would fundamentally transform radio from a temperamental hobby into a reliable mass medium. Yet for all their pioneering importance, crystal receivers possessed inherent limitations that constrained radio's potential: they provided no amplification, required constant adjustment, and could only receive the strongest local signals with acceptable quality. The delicate cat's whisker detectors that made crystal sets possible were maddeningly inconsistent, losing their sensitivity at the slightest vibration and requiring endless tweaking to maintain even marginal performance.

Into this world of whispered signals and frustrated listeners came a revolution in glass and tungsten that would change everything: the vacuum tube. These glowing electronic valves, evolved from Thomas Edison's incandescent light bulbs, possessed an almost magical ability to amplify weak signals hundreds or thousands of times, transforming the barely audible whispers received by crystal sets into powerful signals capable of driving loudspeakers and filling entire rooms with clear, undistorted sound. The vacuum tube revolution of the mid-1920s didn't simply improve radio—it recreated it entirely, enabling the broadcasting boom that would make radio the dominant mass medium of the twentieth century.

The Crystal Set's Beautiful Limitations

To understand the revolutionary impact of vacuum tubes, one must first appreciate the elegant simplicity and fundamental constraints of the crystal receivers that dominated early radio. These devices, marvels of minimalist engineering, operated without any external power source, extracting their energy entirely from the radio waves they received. A crystal set consisted of just four basic components: a tuning coil to select the desired frequency, a variable capacitor to adjust the tuning, a crystal detector to convert radio frequency signals into audio, and headphones to convert the resulting electrical variations into sound waves.

The crystal detector itself was a fascinating device that worked through principles not fully understood until the development of semiconductor theory decades later. A small piece of galena, pyrite, or other mineral crystal was mounted in a brass cup and contacted by an extremely fine wire called a cat's whisker. This crystal-metal contact created what we now recognize as a primitive semiconductor diode, allowing electrical current to flow in only one direction and thereby rectifying the amplitude-modulated radio signals to recover the audio information they carried.

The performance of crystal sets was both remarkable and frustrating in equal measure. Under ideal conditions, with a good antenna, proper grounding, and atmospheric cooperation, a well-built crystal receiver could detect signals from stations hundreds of miles away. Amateur operators took pride in logging distant stations and comparing reception reports with fellow enthusiasts. The simplicity of crystal sets also made them relatively inexpensive to build and maintain, bringing radio within reach of middle-class families who couldn't afford more sophisticated equipment.

However, crystal sets imposed severe restrictions on radio's practical applications. The complete absence of amplification meant that only relatively strong signals could be received with acceptable quality, limiting most listeners to local or regional stations during daylight hours. The crystal detector required frequent adjustment as the cat's whisker lost contact with sensitive spots on the crystal surface, a process that demanded both patience and delicate touch. Reception quality varied dramatically with atmospheric conditions, antenna configuration, and even the time of day, making consistent performance nearly impossible to achieve.

Perhaps most significantly, crystal sets required headphones for listening, making radio an inherently solitary activity. While family members could take turns listening to popular programs, they couldn't share the experience simultaneously. This limitation prevented radio from becoming the family gathering point it would later become and restricted its potential as a mass medium for home entertainment.

Edison's Neglected Discovery

The technological foundation for vacuum tube amplification had been established decades before anyone recognized its potential for radio applications. In 1883, while working to improve the performance of his incandescent light bulbs, Thomas Edison made a curious discovery that would later prove crucial to the development of electronics. He observed that electrical current could flow through the vacuum inside a light bulb from a heated filament to a cold metal plate, but only in one direction.

Edison patented this phenomenon, which became known as the "Edison effect," but he saw no practical application for it and focused his attention on more immediately profitable inventions. The Edison effect remained a scientific curiosity for nearly two decades, occasionally mentioned in physics textbooks but ignored by inventors and engineers who were busy developing the practical electrical technologies—motors, generators, lighting systems, and power distribution networks—that were transforming industrial society.

The scientific explanation for the Edison effect wouldn't be fully understood until the development of atomic theory in the early twentieth century. The heated filament in Edison's bulbs was actually emitting electrons through a process called thermionic emission—when metal is heated to sufficiently high temperatures, some electrons gain enough energy to escape from the metal's surface and travel through the surrounding vacuum. A positively charged electrode placed in this electron stream would attract the electrons, creating a net flow of electrical current through the vacuum.

This thermionic emission process was inherently directional. Electrons flowed readily from the hot filament to a cold electrode, but no corresponding flow occurred in the reverse direction because the cold electrode emitted no electrons of its own. This asymmetrical behavior made Edison's vacuum bulb behave like an electrical valve that allowed current to pass in only one direction—precisely the property needed for rectifying alternating current signals and detecting amplitude-modulated radio transmissions.

The potential applications of the Edison effect became apparent only when inventors began looking for better ways to detect radio signals. The coherers and magnetic detectors used in early wireless systems were insensitive and unreliable, creating bottlenecks that limited the development of practical wireless communication. What was needed was a detector that could respond reliably to weak signals while providing consistent performance under varying operating conditions.

Fleming's Valve: The First Practical Vacuum Tube

The transformation of Edison's scientific curiosity into a practical electronic device was accomplished by John Ambrose Fleming, a British electrical engineer who had worked with both Edison and Marconi and understood the technical challenges facing wireless communication. In 1904, Fleming realized that the Edison effect could be used to detect radio signals more effectively than any existing method. He developed what he called the "oscillation valve," later known as the Fleming diode or simply the vacuum tube diode.

Fleming's valve consisted of a heated filament (the cathode) and a metal plate (the anode) sealed in an evacuated glass envelope. When radio frequency signals were applied to this device, the unidirectional current flow caused by thermionic emission rectified the alternating current, converting amplitude-modulated radio signals into varying direct current that could drive headphones or other audio devices. The Fleming valve was far more sensitive than crystal detectors and much more reliable than coherers, providing consistent performance that didn't require constant adjustment.

The technical advantages of Fleming's valve were immediately apparent to wireless operators who had struggled with the limitations of earlier detection methods. The vacuum tube detector responded linearly to signal strength, providing clear indication of signal quality and enabling more accurate tuning. It was immune to vibration and mechanical shock, unlike the delicate crystal detectors that could be rendered inoperative by the slightest jarring. Most importantly, the Fleming valve could be optimized for maximum sensitivity by adjusting the filament temperature and plate voltage, providing operator control over receiver performance.

Fleming valves found immediate application in long-distance wireless communication systems where maximum sensitivity was crucial. The Marconi Company began using Fleming valves in its transatlantic wireless stations, where the improved detection capability enabled more reliable communication across the challenging ocean path. Naval and commercial wireless stations also adopted Fleming valves, appreciating their superior performance and reduced maintenance requirements compared to earlier detectors.

However, the Fleming valve, like the crystal detectors it replaced, provided no amplification. It could detect signals more reliably than earlier methods, but it couldn't make weak signals stronger. This limitation meant that Fleming valves, while superior to their predecessors, still couldn't solve the fundamental problem that restricted radio to users with technical expertise and tolerance for marginal performance. The breakthrough that would transform radio into a mass medium required not just better detection, but signal amplification that could boost weak radio signals to levels suitable for driving loudspeakers and serving general audiences.

De Forest's Audion: The Birth of Electronic Amplification

The invention that would revolutionize not only radio but all of electronics came in 1907 when Lee de Forest, working in his New York laboratory, added a third electrode to Fleming's two-element vacuum tube. De Forest's modification was deceptively simple—he inserted a wire mesh grid between the heated filament and the metal plate, creating what he called the "audion" tube. This third electrode, which de Forest initially thought of as a means of improving detection sensitivity, actually created the world's first electronic amplifier.

The audion's amplification capability arose from the unique properties of the electron stream flowing from cathode to anode. The wire mesh grid, positioned in the path of this electron flow, could control the current passing through the tube even when only tiny voltages were applied to it. A small change in grid voltage could cause a much larger change in the current flowing from cathode to anode, providing voltage amplification that could boost weak signals to much higher levels.

De Forest initially didn't fully understand the amplification principle underlying his audion tube. He had developed the device primarily as an improved detector for wireless telegraphy, and it took several years of experimentation before he and other inventors recognized that the audion could amplify signals as well as detect them. This gradual recognition of the audion's amplification capability reflected the revolutionary nature of electronic amplification—it was so fundamentally different from existing technologies that even its inventor needed time to comprehend its full potential.

The audion's amplification mechanism worked through what we now call transconductance—the ability of small grid voltage changes to control much larger anode current variations. When a weak radio signal was applied to the grid of an audion tube, it modulated the electron flow from cathode to anode, creating a much stronger signal in the anode circuit that faithfully reproduced the original signal but at much higher amplitude. Multiple audion tubes could be cascaded, with the output of one tube feeding the input of the next, providing virtually unlimited amplification capability.

The implications of electronic amplification extended far beyond radio applications. The audion made practical long-distance telephony possible, enabling coast-to-coast telephone service in the United States. It also found applications in public address systems, electronic measuring instruments, and countless other devices that required signal amplification. But it was in radio that the audion would have its most dramatic impact, transforming wireless communication from a specialized technical service into a mass entertainment medium.

Early audion tubes were primitive by later standards, with limited amplification capability and relatively short operating lives. The manufacturing processes used to produce them were adapted from light bulb production, resulting in tubes with inconsistent performance and high failure rates. Despite these limitations, even early audion tubes provided enough amplification to make radio signals audible through loudspeakers rather than just headphones, opening the door to family listening and shared radio experiences that would define broadcasting's social impact.

The Westinghouse Revolution

The transformation of vacuum tube technology from laboratory curiosity to mass-market product occurred during the 1920s, driven largely by the Westinghouse Electric Corporation's commitment to radio broadcasting and receiver manufacturing. Westinghouse engineers, led by Frank Conrad and others, recognized that vacuum tube improvements were essential to radio's commercial success and dedicated substantial resources to developing better tubes and more sophisticated receiver circuits.

The technical challenges facing vacuum tube development in the early 1920s were formidable. Manufacturing processes needed to be refined to produce tubes with consistent performance and reasonable cost. Tube designs had to be improved to provide greater amplification, longer operating life, and reduced power consumption. Perhaps most importantly, receiver circuits needed to be developed that could take full advantage of vacuum tube capabilities while remaining simple enough for ordinary consumers to operate and maintain.

Westinghouse's approach to these challenges was characteristically systematic and thorough. The company established dedicated laboratories for vacuum tube research, hired skilled engineers and scientists, and invested heavily in manufacturing equipment designed specifically for tube production. Unlike the lamp manufacturers who had initially produced vacuum tubes as a sideline to their primary business, Westinghouse treated tube development as a core competency essential to their radio broadcasting and receiver manufacturing strategies.

The technical improvements achieved by Westinghouse and other manufacturers during the mid-1920s were remarkable in both scope and impact. Tube amplification factors increased from the modest levels achieved by early audions to much higher values that enabled sensitive receivers with excellent selectivity. Tube reliability improved dramatically as manufacturing processes were refined and quality control procedures were implemented. Operating life increased from hundreds of hours to thousands of hours, making vacuum tube receivers practical for everyday use by ordinary consumers.

Perhaps most significantly, Westinghouse engineers developed more sophisticated receiver circuits that could fully exploit vacuum tube capabilities. These circuits included multiple stages of amplification, improved selectivity through tuned circuits, and better audio quality through careful attention to frequency response and distortion characteristics. The result was a new generation of radio receivers that performed far better than anything previously available while remaining simple enough for non-technical users to operate successfully.

The Battery Eliminator Revolution

One of the most significant barriers to radio's mass adoption was the complex power supply requirements of early vacuum tube receivers. The tubes required two different types of electrical power: alternating current to heat the filaments and direct current at various voltages to operate the amplifier circuits. Early receivers used batteries to supply these power requirements, creating ongoing expense and maintenance problems that limited radio's appeal to general audiences.

The battery requirements for vacuum tube receivers were both complex and costly. The filament batteries, which had to supply several amperes of current to heat multiple tubes, needed frequent replacement or recharging. The high-voltage batteries that supplied anode power were expensive and had limited life, especially when used with powerful receivers. Many receivers also required a separate bias battery to provide the proper grid voltage for optimal tube operation. Managing all these batteries required technical knowledge and imposed ongoing costs that made radio ownership burdensome for many families.

The solution to the battery problem came in 1926 with the introduction of the "battery eliminator," a device that could convert household alternating current into the direct current voltages required by vacuum tube receivers. Battery eliminators used transformers to step household voltage up or down to the levels required by different receiver circuits, then employed rectifier tubes to convert alternating current to direct current. Filter circuits smoothed the rectified voltage to provide the steady direct current that vacuum tubes required for optimal operation.

The introduction of battery eliminators transformed radio from a hobby requiring technical maintenance into an appliance as simple to operate as an electric light. Listeners could now enjoy radio without worrying about battery condition, replacement costs, or the technical complexities of maintaining multiple power supplies. The convenience of plug-in operation made radio accessible to families who had been deterred by the maintenance requirements of battery-powered receivers.

However, early battery eliminators still required separate filament batteries because the rectifier tubes available in 1926 couldn't efficiently supply the high current needed to heat receiver tube filaments. This limitation was overcome with the development of indirectly heated vacuum tubes, which used separate heater elements that could operate on alternating current while maintaining the cathode at a steady potential. By 1927, the first completely battery-free radio receivers became available, marking the final step in radio's evolution from technical curiosity to consumer appliance.

The Superheterodyne Revolution

While vacuum tubes solved radio's amplification problems, they couldn't address all the limitations that prevented radio from reaching its full potential as a mass medium. Early vacuum tube receivers, while much more sensitive than crystal sets, still suffered from poor selectivity—the ability to separate signals from different stations operating on nearby frequencies. This limitation became increasingly problematic as the number of broadcasting stations multiplied in the early 1920s, creating interference problems that degraded reception quality and frustrated listeners.

The solution to radio's selectivity problem came from an unlikely source: the French military's efforts to intercept German radio communications during World War I. French engineer Lucien Lévy had developed a receiver design that could intercept enemy signals with exceptional selectivity and sensitivity, but the technique remained classified until after the war. In America, Edwin Armstrong independently developed a similar approach that he called the "superheterodyne" principle, which would revolutionize radio receiver design.

The superheterodyne principle worked by converting incoming radio frequency signals to a fixed intermediate frequency where amplification and filtering could be performed more effectively. Instead of trying to amplify and filter signals at their original broadcast frequencies—which varied across the radio spectrum—superheterodyne receivers converted all signals to the same intermediate frequency, typically 455 kilohertz. This frequency conversion allowed receiver designers to optimize amplifier and filter circuits for maximum performance at a single frequency rather than trying to achieve good performance across the entire broadcast band.

The technical advantages of superheterodyne operation were immediately apparent to anyone who compared superheterodyne receivers with earlier designs. Selectivity was dramatically improved, allowing listeners to separate stations operating on adjacent frequencies without interference. Sensitivity was enhanced because the intermediate frequency amplifier stages could be optimized for maximum gain. Most importantly, tuning became much easier because all the receiver's circuits could be ganged together and controlled with a single knob, eliminating the complex multi-dial tuning procedures required by earlier receivers.

The first commercial superheterodyne receivers appeared in 1924, but their high cost initially limited their market to wealthy enthusiasts willing to pay premium prices for superior performance. The RCA Radiola III, one of the first mass-produced superheterodyne receivers, sold for $286—equivalent to several thousand dollars in modern currency. This pricing reflected both the complexity of superheterodyne circuits and the premium that early adopters were willing to pay for the latest technology.

Cost reduction came gradually as manufacturers developed more efficient production methods and component suppliers achieved economies of scale. By 1929, Radiola 66 and Radiola 67 superheterodyne receivers were being offered at prices that brought superheterodyne performance within reach of middle-class families. These receivers represented the culmination of 1920s radio technology, combining vacuum tube amplification, battery eliminator power supplies, and superheterodyne selectivity in packages designed for easy operation by non-technical users.

Manufacturing and Mass Production

The transformation of vacuum tubes from expensive laboratory devices to mass-market products required fundamental changes in manufacturing processes and business organization. The small-scale production methods that had been adequate for supplying the limited amateur radio market couldn't meet the enormous demand created by the broadcasting boom of the 1920s. New manufacturing techniques, quality control procedures, and distribution systems had to be developed to support radio's growth from hobby to mass medium.

Vacuum tube manufacturing in the early 1920s was still largely a craft industry, with skilled workers hand-assembling tubes using techniques adapted from light bulb production. Each tube was individually constructed, tested, and adjusted, resulting in high costs and inconsistent performance. The manufacturing yields were low, with many tubes failing quality tests and having to be discarded. This hand-crafted approach couldn't scale to meet the millions of tubes needed for mass market radio production.

The solution came through the application of mass production principles that had revolutionized automobile manufacturing. Companies like RCA, Westinghouse, and General Electric invested heavily in automated production equipment that could manufacture tubes with greater consistency and much lower cost. Assembly line techniques were adapted to tube production, with workers specializing in specific operations rather than building complete tubes individually. Quality control procedures were implemented to ensure that tubes met performance standards before leaving the factory.

The development of specialized tube manufacturing equipment was crucial to achieving the cost reductions needed for mass market success. Automatic glass-working machines could form tube envelopes with greater precision and consistency than hand methods. Automatic assembly equipment could install tube elements with exact positioning and proper spacing. Most importantly, new vacuum pumping systems could evacuate large numbers of tubes simultaneously, dramatically reducing the time and cost required for this critical manufacturing step.

Material improvements also contributed to better tube performance and lower costs. The development of oxide-coated cathodes reduced the operating temperature required for thermionic emission, allowing closer spacing of tube elements and improving amplification performance. Better glass formulations provided tube envelopes that were stronger and more resistant to thermal stress. Improved metal alloys for grids and anodes enhanced tube reliability and extended operating life.

The result of these manufacturing improvements was a dramatic reduction in tube costs and a corresponding improvement in performance and reliability. Tubes that had cost dollars in the early

1920s were available for pennies by the end of the decade. Performance improvements meant that fewer tubes were needed to achieve the same results, while reliability improvements reduced replacement costs and maintenance requirements. These advances made vacuum tube radio receivers accessible to ordinary consumers and enabled the mass market that broadcasting required.

Circuit Innovation and Receiver Design

The availability of reliable, inexpensive vacuum tubes enabled radio engineers to develop increasingly sophisticated receiver circuits that took full advantage of vacuum tube capabilities. Early vacuum tube receivers had been little more than crystal sets with vacuum tube detectors substituted for crystal detectors, but by the mid-1920s, entirely new receiver architectures were emerging that exploited vacuum tubes' unique properties to achieve performance levels impossible with earlier technologies.

The development of multi-stage amplification represented one of the most important advances in vacuum tube receiver design. Early receivers typically used a single vacuum tube for detection, sometimes followed by one additional tube for audio amplification. Multi-stage receivers used separate tubes for radio frequency amplification, detection, intermediate frequency amplification (in superheterodyne designs), and audio amplification, providing much greater overall sensitivity and better signal quality.

Tuned radio frequency (TRF) receivers represented an intermediate step between simple crystal sets and sophisticated superheterodyne designs. TRF receivers used multiple stages of radio frequency amplification, each tuned to the desired signal frequency, followed by detection and audio amplification stages. While more complex than simple receivers, TRF designs avoided the frequency conversion circuitry required by superheterodyne receivers, making them somewhat less expensive to manufacture. However, TRF receivers suffered from alignment difficulties and limited selectivity that made them inferior to superheterodyne designs for demanding applications.

The development of automatic gain control (AGC) circuits represented another significant advance in receiver technology. Early receivers provided manual volume controls that listeners had to adjust constantly to compensate for signal strength variations caused by fading and atmospheric conditions. AGC circuits automatically adjusted receiver gain to maintain consistent output volume despite input signal variations, greatly improving the listening experience and making radio more convenient for casual users.

Tone control circuits were developed to compensate for the frequency response limitations of early audio systems and to allow listeners to adjust sound quality to their preferences. These circuits, typically consisting of simple high-frequency and low-frequency filters, enabled listeners to emphasize or de-emphasize different parts of the audio spectrum. While primitive by modern standards, early tone controls represented an important step toward high-fidelity audio reproduction.

The integration of multiple functions into single vacuum tubes represented another important development in receiver design. Combination tubes that performed both detection and amplification functions reduced tube count and simplified receiver circuits while maintaining good performance. These multi-function tubes were particularly important for portable and automotive receivers where space and power consumption were critical considerations.

The Impact on Broadcasting Quality

The vacuum tube revolution didn't only transform radio receivers—it also enabled dramatic improvements in broadcasting equipment that enhanced the quality and reliability of transmitted programs. Early broadcasting stations had used simple vacuum tube circuits for modulation and amplification, but as tube technology matured, station engineers developed increasingly sophisticated transmission systems that could deliver high-quality audio with minimal distortion and maximum reliability.

Transmitter power levels increased substantially as more powerful vacuum tubes became available. Early broadcasting stations typically operated with powers of a few hundred watts, limiting their coverage areas to local regions. By the late 1920s, stations were using kilowatt-level transmitters that could provide reliable coverage over entire states or regions. This power increase was made possible by the development of high-power vacuum tubes that could handle the current and voltage levels required for effective radio transmission.

Audio quality improvements were equally significant, as station engineers learned to design circuits that minimized distortion and maximized frequency response. Early broadcast transmissions had been characterized by high levels of harmonic distortion and limited frequency response that made speech difficult to understand and music unpleasant to hear. Improved vacuum tube circuits, better modulation techniques, and more sophisticated audio processing equipment gradually raised broadcast audio quality to levels that approached the fidelity of high-quality phonograph reproduction.

The development of remote broadcasting capability was another important advance enabled by vacuum tube technology. Early broadcasting stations were limited to programming that could be produced in their studios, but vacuum tube amplification made it practical to broadcast live performances from concert halls, theaters, sporting events, and other remote locations. High-quality telephone lines could carry audio signals from remote locations to broadcasting stations, where vacuum tube amplifiers boosted the signals to levels suitable for transmission.

Studio equipment also benefited from vacuum tube advances, with improved microphones, mixing equipment, and monitoring systems that enabled more professional program production. Multiple microphone inputs could be mixed together to create complex programs featuring multiple performers, while monitoring equipment allowed engineers to ensure optimal audio quality throughout the broadcast chain. These technical improvements elevated broadcasting from amateur experimentation to professional entertainment production that could compete with other media for audience attention.

The Economic Transformation

The vacuum tube revolution transformed radio from an expensive hobby pursued by technical enthusiasts into a mass consumer market that supported entire industries. The improved performance and reduced costs enabled by vacuum tube technology created the market conditions necessary for radio manufacturing to become a major economic sector, employing thousands of workers and generating millions of dollars in revenue.

Radio receiver sales exploded as vacuum tube technology made radio accessible to general audiences. In 1920, fewer than 50,000 households owned radio receivers; by 1930, over 12 million American families had radios, representing one of the fastest technology adoption rates in history. This growth was driven primarily by the performance improvements and cost reductions made possible by vacuum tube advances, which transformed radio from a difficult hobby into an convenient entertainment appliance.

The vacuum tube industry itself became a major economic sector, with companies like RCA, General Electric, and Westinghouse investing heavily in tube production facilities and research laboratories. Tube sales generated substantial revenue streams that funded continued research and development, creating a virtuous cycle of technological improvement and market expansion. The employment generated by tube manufacturing, receiver production, and broadcasting operations contributed significantly to the prosperity of the 1920s.

Component suppliers also benefited from radio's growth, with companies specializing in capacitors, resistors, transformers, and other radio components experiencing unprecedented demand. The standardization that accompanied mass production enabled these suppliers to achieve economies of scale that reduced costs and improved quality. The entire electronics industry ecosystem that emerged during the 1920s was built on the foundation of vacuum tube technology and the mass market it created.

The advertising industry found new opportunities in radio broadcasting, as vacuum tube receivers enabled the high-quality audio reproduction necessary for effective commercial messages. Radio advertising revenue provided the economic foundation that supported program production and station operations, creating the business model that sustained commercial broadcasting throughout the twentieth century. The combination of technical capability and economic incentive created by vacuum tube technology established the framework for modern mass media.

The Social Revolution Continues

The technical improvements enabled by vacuum tubes had profound social implications that extended far beyond their immediate impact on radio performance. By making radio accessible to ordinary families and enabling high-quality audio reproduction through loudspeakers, vacuum tubes transformed radio from an individual activity requiring headphones into a shared family experience that could bring households together around common entertainment and information sources.

The ability to receive distant stations clearly and consistently expanded listeners' horizons beyond their immediate communities, connecting rural areas to urban culture and breaking down the geographic isolation that had historically characterized American life. Farmers in remote areas could receive the same entertainment programs enjoyed by city dwellers, while urban listeners could enjoy country music and agricultural programming that connected them to rural America. This cultural mixing contributed to the development of more homogeneous national culture and shared experiences that transcended regional boundaries.

Educational opportunities were also enhanced by vacuum tube technology, as clear reception enabled effective distance learning through radio broadcasts. Universities began offering courses by radio, while public broadcasting efforts brought lectures, concerts, and cultural programming into homes across America. The improved audio quality made possible by vacuum tubes was essential to these educational applications, as speech clarity and music fidelity were crucial for effective educational communication.

Emergency communication capabilities were dramatically enhanced by vacuum tube receivers that could receive weak distant signals clearly and consistently. During natural disasters and other emergencies, radio became the primary means of coordinating relief efforts and maintaining communication with isolated areas. The reliability and sensitivity of vacuum tube receivers made them essential tools for emergency preparedness and disaster response.

The Foundation for Future Innovation

The vacuum tube revolution of the 1920s established the technological foundation for virtually every subsequent development in electronics and communications. The principles of electronic amplification, signal processing, and circuit design that were pioneered in vacuum tube radio systems became the building blocks for television, radar, computers, and countless other electronic technologies that would define the twentieth century.

The manufacturing techniques developed for vacuum tube production provided the basis for the mass production of electronic components that would later enable the transistor revolution and the development of integrated circuits. The quality control procedures, testing methods, and reliability engineering practices developed for vacuum tubes established the standards and practices that would guide the electronics industry throughout its subsequent evolution.

Perhaps most importantly, vacuum tubes demonstrated the transformative power of electronic technology to reshape society and create new forms of human communication and entertainment. The social changes brought about by vacuum tube radio—the development of mass media, the creation of shared national culture, the democratization of entertainment and information—prefigured even greater transformations that would accompany television, computers, and digital communication technologies.

The glowing tubes that had revolutionized radio in the 1920s would eventually be superseded by transistors and integrated circuits, but their legacy lived on in the electronic civilization they had made possible. The technical leap represented by vacuum tubes had indeed changed everything, creating the mass communication society that would define the modern world. Radio had found its voice through vacuum tubes, and that voice would speak to humanity for generations to come, carrying news and entertainment, education and culture, across the invisible waves that connected an increasingly united world.

Chapter 7: Around the World on Shortwave

In the summer of 1923, as jazz music filled the airwaves of America's new broadcasting stations and families gathered around crystal sets in living rooms across the continent, a quiet revolution was unfolding on the windswept coast of Cornwall, England. At the Poldhu Wireless Station—the same facility from which Guglielmo Marconi had transmitted his first transatlantic signals over two decades earlier—a new chapter in radio's story was being written, one that would ultimately shrink the world and connect humanity across vast oceans and continents through the remarkable propagation characteristics of shortwave radio.

The man at the center of this revolution was Charles Samuel Franklin, Marconi's brilliant assistant, who had been commissioned to conduct what would prove to be one of the most significant investigations in radio history. Working with a massive 25-kilowatt transmitter and experimental antenna arrays, Franklin was exploring the transmission characteristics of short-wavelength radio waves that most experts had dismissed as useless for long-distance communication. The conventional wisdom of the early 1920s held that reliable long-distance radio communication required the massive power and enormous antennas associated with longwave transmission—wavelengths of thousands of meters that could follow the curvature of the Earth and bridge continents through ground wave propagation.

But Franklin's experiments would shatter this conventional wisdom and reveal that nature had provided radio engineers with an extraordinary gift: the ionosphere, a layer of electrically charged particles in the upper atmosphere that could reflect shortwave signals back to Earth, enabling global communication with relatively modest power and compact antennas. This discovery would not only revolutionize international communication but would also create the technical foundation for international broadcasting, amateur radio's global networks, and emergency communications that could span the globe when all other systems failed.

The Great Shortwave Discovery

The story of shortwave radio's emergence as a practical communication medium begins with a fundamental misconception that had dominated radio engineering since the earliest days of wireless communication. Prior to the 1920s, radio engineers believed that reliable long-distance communication required wavelengths below 300 kilohertz—the longwave spectrum that extended beyond 1,000 meters. These extremely long wavelengths could propagate around the Earth's curvature through ground wave transmission, but they required enormous transmitter powers, massive antenna installations, and generated signals so broad that only a few stations could operate simultaneously without interfering with each other.

The shortwave frequencies above 1.5 MHz, corresponding to wavelengths shorter than 200 meters, were regarded as essentially worthless for serious communication applications. These frequencies had been allocated to amateur radio operators in many countries precisely because professional radio engineers considered them too limited for commercial or military use. The prevailing theory suggested that these short wavelengths would travel only in straight lines, limiting their range to the radio horizon—perhaps 50 to 100 miles under the best conditions.

Franklin's systematic investigation of shortwave propagation characteristics began in June 1923 with experimental transmissions on 97 meters (approximately 3 MHz) from the Poldhu station to Marconi's yacht Elettra in the Cape Verde Islands. The initial results were encouraging but not revolutionary—the 97-meter signals could be received reliably during nighttime hours, suggesting that shortwave might have some utility for long-distance communication when atmospheric conditions were favorable.

But it was the experiments conducted in September 1924 that truly changed everything. Franklin arranged for continuous transmissions on 32 meters (approximately 9.4 MHz) from Poldhu to the Elettra, which had sailed to Beirut in the eastern Mediterranean. Marconi himself was aboard the yacht to witness the results, and what he observed left him "astonished"—the shortwave signals could be received clearly not just during the nighttime hours when such reception might be expected, but "throughout the day" with consistent strength and clarity.

This daylight reception of shortwave signals across thousands of miles contradicted everything that radio engineers thought they knew about radio wave propagation. The signals were not traveling along the Earth's surface as ground waves, nor were they limited to line-of-sight distances as conventional theory predicted. Instead, they were somehow being reflected or refracted by something in the upper atmosphere, bouncing between the Earth and the sky to reach distant receivers with remarkable efficiency.

The implications of this discovery were staggering. If shortwave signals could reliably reach distant continents during both day and night, then the massive longwave installations that dominated international communication might be unnecessary. Shortwave stations could provide global coverage with a fraction of the power, antenna size, and cost required for longwave systems. This promised to democratize international communication and make global radio networks economically feasible for the first time.

The Imperial Wireless Chain

Recognizing the revolutionary potential of Franklin's discoveries, Marconi moved quickly to transform shortwave research into commercial reality. In July 1924, even before the full implications of the 32-meter experiments were understood, Marconi entered into contracts with the British General Post Office to establish high-speed shortwave telegraphy circuits connecting London to Australia, India, South Africa, and Canada. This ambitious project, known as the Imperial Wireless Chain, would demonstrate shortwave radio's capability to serve as the backbone of a global communication network spanning the British Empire.

The Imperial Wireless Chain represented one of the most significant technical and commercial achievements in radio history. Unlike the point-to-point longwave circuits that had previously connected Britain to its colonies, the shortwave system could provide multiple communication channels simultaneously while operating at a fraction of the cost. The UK-to-Canada shortwave "Beam Wireless Service" began commercial operation on October 25, 1926, followed by services to Australia, South Africa, and India in 1927.

The technical challenges of establishing reliable shortwave communication across such vast distances were formidable. Franklin developed sophisticated directional antenna systems, including his innovative curtain array design, that could focus shortwave signals toward their intended destinations while minimizing interference to other services. These directional antennas were crucial to shortwave's success, as they concentrated the transmitted power in the desired direction while reducing the massive power requirements that had made longwave systems so expensive.

The economic impact of the Imperial Wireless Chain was immediate and dramatic. The efficient shortwave circuits provided communication services at rates far below those charged by the submarine cable companies that had previously monopolized international communication. By 1928, more than half of all long-distance communications had migrated from transoceanic cables and longwave services to shortwave systems. The volume of international communication increased dramatically as the lower costs made global communication accessible to businesses and organizations that had previously been priced out of the market.

However, the success of shortwave communication also created a serious crisis for the cable companies that had invested enormous sums in submarine telegraph cables. These companies began losing large amounts of money in 1927 as customers switched to the cheaper and more flexible shortwave services. The financial crisis was so severe that it threatened the viability of cable companies that were considered vital to British strategic interests, leading to government intervention and the eventual merger of cable and wireless resources into a single company, Cable and Wireless Ltd., in 1934.

The Amateur Radio Revolution

While Marconi and his commercial successors were developing shortwave for imperial communication networks, a parallel revolution was unfolding in the amateur radio community. The radio amateurs who had been relegated to the "useless" shortwave frequencies were discovering that these wavelengths possessed extraordinary capabilities for long-distance communication that far exceeded anything possible with their previous longwave allocations.

The transformation of amateur radio from a local hobby to a global phenomenon began in the early 1920s as experimenters discovered that shortwave signals could propagate across continents and oceans with relatively simple equipment. Prior to 1920, amateur radio operators had been restricted to wavelengths of 200 meters and below (1,500 kHz and above), which had severely limited their communication range to perhaps 25 miles under typical conditions. This limitation had required the American Radio Relay League (ARRL) to establish elaborate relay networks to carry messages across long distances, with each amateur station passing messages to the next station in the chain.

The Radio Act of 1912 had further restricted amateur operations by designating the 200-meter wavelength as the upper limit for "private stations," reflecting the government's belief that these shorter wavelengths were unsuitable for serious communication purposes. The number of active amateur stations had dropped from an estimated 10,000 to just 1,200 following the implementation of these restrictions, as many operators found the limited range and interference problems of the 200-meter band too frustrating to maintain their interest in radio.

But the situation changed dramatically in 1924 when amateurs were allocated new bands at 80, 40, 20, and 5 meters (corresponding to frequencies of approximately 3.5, 7, 14, and 60 MHz). These allocations reflected growing recognition that shortwave frequencies could provide valuable communication capabilities, while the prohibition of spark transmitters on the new bands encouraged the development of continuous wave (CW) transmissions that were more efficient and produced less interference.

The results exceeded everyone's expectations. By 1925, amateur radio operators were successfully communicating around the world using shortwave frequencies, achieving reliable contact between stations separated by thousands of miles. The amateur radio census reached 14,000 operators by 1923 and continued growing rapidly as word spread about shortwave's remarkable capabilities. The formation of the International Amateur Radio Union (IARU) in 1925 reflected the global nature of amateur radio communication and the need for international coordination of amateur activities.

Amateur radio's success with shortwave communication had profound implications that extended far beyond the hobby itself. Amateur operators served as an informal research and development laboratory for shortwave techniques, experimenting with new antenna designs, modulation methods, and operating procedures that advanced the entire field of shortwave communication. Their experiences provided valuable data about shortwave propagation characteristics and demonstrated the reliability of shortwave communication under diverse conditions.

Early International Broadcasting Experiments

The success of shortwave for point-to-point communication naturally led to experiments with international broadcasting—the transmission of programs intended for general audiences in distant countries rather than for specific receiving stations. These early experiments in international broadcasting faced unique challenges that differed significantly from domestic broadcasting or commercial point-to-point communication.

Unlike domestic broadcasting, which served audiences within a relatively small geographic area with predictable propagation characteristics, international broadcasting had to overcome the variability of shortwave propagation across different time zones, seasons, and atmospheric conditions. The same frequency that provided excellent coverage of Europe during winter evenings might be completely unusable for the same target area during summer afternoons. This variability required international broadcasters to develop sophisticated frequency management strategies and to operate multiple transmitters on different frequencies simultaneously.

The technical requirements for international broadcasting also differed from those of commercial communication services. While point-to-point circuits could tolerate some signal distortion and noise as long as the message content remained intelligible, broadcasting required higher audio quality to ensure that speech was clearly understandable and music was pleasant to hear. This requirement was particularly challenging for shortwave transmission, which was subject to fading, interference, and atmospheric noise that could severely degrade audio quality.

Despite these challenges, experimental international broadcasts began appearing in the mid-1920s as various organizations recognized shortwave's potential for reaching international audiences. Religious organizations were among the early pioneers, using shortwave to reach missionaries and congregations in remote areas where no other communication medium was available. Educational institutions began broadcasting lectures and cultural programs to international audiences, while commercial broadcasters experimented with shortwave transmission to extend their coverage areas beyond their domestic markets.

The content of early international broadcasts reflected both the technical limitations of shortwave transmission and the cultural assumptions of the broadcasting organizations. Programs were typically simple in format, emphasizing clear speech over complex musical arrangements that might not survive shortwave transmission intact. News programming became particularly important for international broadcasting, as shortwave's ability to provide real-time information across vast distances gave it significant advantages over newspapers and magazines for reaching international audiences.

The BBC World Service Pioneer

The British Broadcasting Corporation's entry into international shortwave broadcasting marked a crucial turning point in the development of shortwave as a medium for international communication and cultural exchange. The BBC's approach to international broadcasting was systematic and well-funded, reflecting Britain's recognition that shortwave radio could serve important political and cultural objectives in maintaining connections with the British Empire and projecting British influence around the world.

The BBC's international shortwave service, which would eventually become the BBC World Service, began with experimental transmissions in the early 1930s. Unlike the amateur experiments and commercial ventures that had preceded it, the BBC brought professional broadcasting expertise, substantial financial resources, and a clear organizational mission to international shortwave broadcasting. The corporation's domestic broadcasting experience provided valuable insights into program production, audience research, and technical standards that could be adapted for international shortwave transmission.

The BBC's approach to international programming reflected a sophisticated understanding of both the technical capabilities and limitations of shortwave transmission. Programs were designed specifically for shortwave transmission, taking into account the fading and interference characteristics that could affect signal quality. Speech programs emphasized clear diction and slower delivery rates to ensure intelligibility under difficult reception conditions, while musical selections were chosen for their suitability to shortwave transmission characteristics.

The geographic scope of BBC shortwave broadcasting was ambitious from the beginning, reflecting Britain's global interests and the technical capabilities of shortwave propagation. Different program services were developed for different regions of the world, with transmission schedules and frequency selections optimized for the propagation characteristics of each target area. This regional approach recognized that effective international broadcasting required more than simply transmitting domestic programs to international audiences—it demanded understanding of local cultures, languages, and interests in the target areas.

The BBC's commitment to high technical standards had important implications for the entire field of international shortwave broadcasting. The corporation invested heavily in high-power transmitters, sophisticated antenna systems, and professional studio facilities that established new benchmarks for shortwave broadcast quality. These technical improvements demonstrated that shortwave could provide audio quality approaching that of domestic medium wave broadcasting when sufficient resources were applied to the challenge.

The Golden Age Emerges

The period from the 1930s through the 1960s would later be recognized as the golden age of international shortwave broadcasting, when hundreds of stations around the world competed for listeners' attention and shortwave receivers in millions of homes provided windows to the world that were unmatched by any other communication medium. This golden age was characterized by rapid growth in both the number of international broadcasting stations and the sophistication of their programming and technical capabilities.

The political tensions of the 1930s provided strong motivation for governments to establish international shortwave broadcasting services. As international relations deteriorated and the threat of global conflict increased, nations recognized that shortwave radio offered an important means of communicating with their citizens abroad, influencing international opinion, and gathering intelligence about foreign activities. The relatively low cost of establishing shortwave broadcasting services made them accessible to even modest national budgets, while their global reach provided influence capabilities that had previously been available only to major powers with worldwide communication networks.

The technical infrastructure supporting international shortwave broadcasting expanded rapidly during this period. High-power transmitters capable of generating hundreds of kilowatts became available, providing the signal strength needed for reliable global coverage. Sophisticated antenna systems, including rotatable arrays and electronically steered beams, enabled broadcasters to focus their signals toward specific target areas while minimizing interference to other services. Multiple frequency operations became standard, with stations broadcasting simultaneously on several frequencies to ensure reliable reception despite propagation variations.

Programming diversity flourished during the golden age as broadcasters competed for international audiences. News services became increasingly important, with stations providing detailed coverage of international events from their national perspectives. Cultural programming showcased each country's music, literature, and traditions, while educational services offered language instruction and technical training to international audiences. Entertainment programming adapted domestic formats for international transmission, creating new forms of radio drama and variety shows designed specifically for shortwave audiences.

The listener experience during the golden age was unlike anything available through other media. Shortwave listeners could explore the world from their homes, tuning across the spectrum to hear news from

Moscow, music from Radio Australia, religious programming from the Vatican, and amateur radio conversations from stations around the globe. This diversity of content and the unpredictable nature of shortwave propagation created a sense of adventure and discovery that made shortwave listening a popular hobby in its own right.

Reception technology also advanced significantly during the golden age, with manufacturers developing specialized shortwave receivers that could handle the unique requirements of international broadcasting reception. These receivers featured multiple conversion systems for improved sensitivity and selectivity, bandspread tuning for precise frequency control, and automatic gain control to compensate for signal fading. The availability of high-quality shortwave receivers made international broadcasting accessible to general audiences rather than just technical specialists.

Emergency Communications and Global Networks

One of shortwave radio's most valuable contributions to human welfare has been its role in emergency communications when other communication systems fail or are unavailable. The same propagation characteristics that make shortwave ideal for international broadcasting—the ability to span continents and oceans with relatively simple equipment—also make it invaluable for coordinating disaster relief, maintaining communication with remote areas, and providing backup communication capabilities when terrestrial networks are disrupted.

Amateur radio operators have played a particularly important role in developing and maintaining emergency communication capabilities using shortwave frequencies. The amateur radio community's tradition of experimentation and technical innovation, combined with their geographic distribution and dedication to public service, has created an informal but highly effective global emergency communication network that can be activated within hours of a disaster.

The value of amateur radio for emergency communication was demonstrated repeatedly throughout the golden age of shortwave broadcasting. Arctic and Antarctic expeditions, including those led by Commander Byrd, relied on amateur radio operators and shortwave equipment as their primary means of maintaining contact with civilization. The MacMillan Arctic Expedition was among the first to carry two-way radio equipment, using an amateur 200-meter station to maintain communication links that would have been impossible with any other technology available at the time.

Natural disasters provided frequent demonstrations of shortwave radio's emergency communication capabilities. When earthquakes, hurricanes, floods, or other disasters disrupted telephone and telegraph services, amateur radio operators could often establish communication links within hours using portable shortwave equipment powered by batteries or emergency generators. These emergency networks provided crucial coordination capabilities for relief operations and enabled families to learn about the safety of relatives in affected areas.

The international nature of shortwave propagation made it particularly valuable for coordinating international relief efforts and maintaining communication with disaster areas that were isolated from their own national communication networks. Amateur radio operators in unaffected areas could relay messages between disaster zones and relief organizations, providing communication links that spanned continents and enabled rapid coordination of international assistance.

Military and maritime applications also benefited from shortwave radio's emergency communication capabilities. Ships at sea could maintain contact with shore stations and other vessels using shortwave equipment that was much more compact and economical than the longwave systems previously required for long-distance maritime communication. Military forces found that shortwave provided reliable communication capabilities under conditions where other systems might be compromised or unavailable.

Propagation Science and Understanding

The practical success of shortwave communication stimulated intensive scientific research into the propagation characteristics that made long-distance shortwave transmission possible. Understanding these propagation mechanisms was crucial for optimizing shortwave communication systems and predicting the performance of different frequencies under various conditions.

The key to understanding shortwave propagation lay in recognizing the role of the ionosphere—layers of electrically charged particles in the upper atmosphere that could reflect radio waves back to Earth. Solar radiation ionizes gases in the upper atmosphere, creating regions of free electrons and ions that interact with radio waves passing through them. Under the right conditions, these ionized layers act like mirrors in the sky, reflecting shortwave signals back to Earth at distances far beyond the radio horizon.

The ionosphere's behavior varies with solar activity, time of day, season, and geographic location, creating the complex propagation patterns that characterize shortwave communication. Different frequencies are affected differently by ionospheric conditions, with lower frequencies generally providing better propagation during nighttime hours and higher frequencies working better during daylight. This frequency dependence requires shortwave operators to change frequencies throughout the day to maintain reliable communication links.

Solar activity has a profound effect on shortwave propagation, with the 11-year solar cycle creating long-term variations in ionospheric conditions. High solar activity generally improves propagation on higher shortwave frequencies while potentially disrupting communication through magnetic storms and other phenomena. Understanding these solar influences became crucial for frequency planning and propagation prediction, leading to the development of sophisticated forecasting services that could predict shortwave propagation conditions days or weeks in advance.

The scientific understanding of shortwave propagation that developed during the golden age enabled significant improvements in shortwave communication system design. Antenna systems could be optimized for specific propagation paths and frequencies, while transmission schedules could be coordinated with predicted propagation conditions. This scientific approach transformed shortwave from an art based on experience and intuition into an engineering discipline based on theoretical understanding and quantitative prediction.

Cultural Impact and Global Understanding

The golden age of shortwave broadcasting had profound cultural implications that extended far beyond its technical achievements. For the first time in human history, ordinary people could hear directly from other cultures and nations without the filtering and interpretation imposed by their domestic media. This direct cultural contact through shortwave broadcasting contributed to international understanding and created personal connections between peoples that transcended political boundaries.

Shortwave broadcasting provided audiences with access to music, literature, and artistic traditions from around the world. Classical music lovers could hear performances by orchestras from major cultural centers, while folk music enthusiasts could explore traditional music from cultures they might never have encountered otherwise. This cultural exchange enriched listeners' understanding of global diversity and created appreciation for artistic traditions that might otherwise have remained unknown outside their countries of origin.

Educational programming through shortwave broadcasting extended the reach of universities and cultural institutions far beyond their local communities. Language instruction became one of the most popular forms of shortwave educational programming, with services like the BBC's English lessons reaching millions of students worldwide. Technical and vocational training programs helped developing countries access educational resources that were not available through their domestic educational systems.

News and information programming through shortwave broadcasting provided audiences with alternative perspectives on international events that were often unavailable through their domestic media. Listeners could compare different national viewpoints on the same events, developing more nuanced understanding of complex international issues. This exposure to diverse perspectives contributed to the development of more internationally minded populations that were better prepared for the increasingly interconnected world of the late twentieth century.

The shortwave listening hobby that developed during the golden age created an international community of enthusiasts who maintained detailed logs of stations heard, exchanged reception reports with broadcasters, and shared information about propagation conditions and programming schedules. This community fostered international friendships and cultural exchanges that continued long after the golden age of shortwave broadcasting had ended.

Technological Evolution and Innovation

The technical demands of international shortwave broadcasting drove continuous innovation in transmitter design, antenna systems, and auxiliary equipment throughout the golden age. The need to provide reliable global coverage with acceptable audio quality pushed engineers to develop increasingly sophisticated solutions to the challenges posed by shortwave propagation characteristics.

Transmitter power levels increased dramatically during the golden age, with some international broadcasting stations operating transmitters of 500 kilowatts or more. These high-power transmitters required significant advances in vacuum tube technology, cooling systems, and power supply design. The development of specialized shortwave transmitting tubes enabled efficient generation of high-power shortwave signals, while improved modulation systems maintained audio quality despite the high power levels involved.

Antenna technology became increasingly sophisticated as engineers developed systems that could provide optimal coverage of specific geographic areas while minimizing interference to other services. Rotatable antenna arrays enabled broadcasters to redirect their signals toward different target areas throughout the day, following optimal propagation conditions. Electronically steered arrays provided even greater flexibility, allowing rapid changes in beam direction without mechanical rotation.

Frequency management became increasingly complex as the number of shortwave broadcasting stations multiplied and the available spectrum became crowded. International coordination became essential to minimize interference between different services, leading to the development of sophisticated frequency allocation procedures and interference mitigation techniques. The International Telecommunication Union eventually became necessary to coordinate frequency allocations and minimize interference between different services operating in the increasingly crowded shortwave spectrum.

The development of single sideband (SSB) transmission represented one of the most significant technical advances in shortwave communication during the later part of the golden age. SSB transmission, which transmitted only one sideband of an amplitude-modulated signal while suppressing the carrier and the other sideband, provided significant improvements in power efficiency and spectrum utilization. This technique enabled more efficient use of both transmitter power and frequency spectrum, allowing more communication channels to operate in the available bandwidth.

Receiver technology also advanced significantly, with the development of more sophisticated circuits that could handle the demanding requirements of shortwave reception. Crystal-controlled oscillators provided improved frequency stability, while multiple conversion receivers offered better selectivity and image rejection. Automatic gain control systems compensated for the fading characteristics of shortwave signals, while noise limiters and other circuits improved reception quality under difficult conditions.

The Legacy of the Golden Age

The golden age of shortwave broadcasting created a legacy that extended far beyond the technical achievements of the era. The global communication networks established during this period provided the foundation for international understanding and cultural exchange that would prove invaluable during the challenging decades that followed. The technical innovations developed for shortwave communication found applications in countless other fields, from television broadcasting to satellite communications.

The shortwave frequencies that had once been dismissed as worthless became recognized as one of the most valuable parts of the radio spectrum. The propagation characteristics that enabled global communication with relatively simple equipment made shortwave indispensable for applications ranging from international broadcasting to emergency communications. The scientific understanding of shortwave propagation that developed during the golden age provided the foundation for all subsequent developments in high-frequency communication systems.

Perhaps most importantly, the golden age of shortwave broadcasting demonstrated the power of radio waves to connect humanity across the vast distances and cultural barriers that had historically separated different peoples and nations. The invisible waves that carried voices and music around the world created bonds of understanding and shared experience that transcended political boundaries and national differences. As the world moved toward an increasingly interconnected future, the foundations laid during the golden age of shortwave would prove to be among the most enduring contributions to human communication and understanding.

Part III: Spectrum Wars and Technical Evolution (1930-1980)

Chapter 8: Clearer Signals - The FM Revolution

The year 1933 marked a turning point in radio history that would ultimately transform the medium from a world of crackling static and fading signals into the high-fidelity broadcasting we know today. In his modest laboratory at Columbia University, Edwin Howard Armstrong—the same brilliant inventor who had revolutionized radio reception with his superheterodyne circuit—was putting the finishing touches on a revolutionary approach to radio transmission that would challenge everything the broadcasting industry thought it knew about electromagnetic communication. What Armstrong had created was not merely an improvement to existing amplitude modulation (AM) systems, but an entirely new method of encoding information onto radio waves that promised to eliminate the static, interference, and fading that plagued conventional broadcasting.

This new system, which Armstrong called frequency modulation (FM), represented a fundamental philosophical shift in how radio engineers approached the problem of transmitting information through the electromagnetic spectrum. Instead of varying the strength or amplitude of radio waves to carry audio information—the method that had been used since Fessenden's first voice transmissions—FM varied the frequency of the carrier wave itself, creating a signal that was virtually immune to the atmospheric disturbances and electrical interference that made AM broadcasting so frustratingly unreliable. The implications of this breakthrough were staggering: FM promised to deliver radio programming with a clarity and fidelity that had never before been possible, opening the door to high-quality music broadcasting and establishing radio as a serious medium for audiophiles and music lovers.

Yet the path from Armstrong's laboratory breakthrough to commercial FM broadcasting would prove to be one of the most contentious and tragic stories in the history of technology. Despite FM's obvious technical superiority, the established radio industry would resist its adoption for decades, forcing Armstrong to wage a lonely and ultimately ruinous battle to bring his invention to the public. The story of FM's development reveals not only the technical challenges of radio innovation but also the economic and political forces that can delay or distort the introduction of new technologies, even when their benefits are clear and undeniable.

The Static Problem That Plagued AM

To understand the revolutionary nature of Armstrong's FM system, one must first appreciate the fundamental limitations that plagued amplitude modulation broadcasting throughout the 1920s and early 1930s. AM radio, for all its success in creating the broadcasting industry, suffered from an inherent vulnerability to interference that made high-quality audio reproduction virtually impossible under many conditions.

The root of AM's static problem lay in the nature of amplitude modulation itself. In AM transmission, audio information is encoded by varying the strength or amplitude of a high-frequency carrier wave in direct proportion to the audio signal being transmitted. When atmospheric disturbances—such as lightning strikes, electrical storms, or solar activity—create electromagnetic pulses, these pulses add directly to the transmitted signal, appearing as loud pops, crackles, and hisses in the receiver's output. Since both the desired audio information and the unwanted atmospheric noise were encoded as amplitude variations, there was no way for a receiver to distinguish between the two.

This fundamental limitation meant that AM broadcasting was perpetually at the mercy of atmospheric conditions. Summer thunderstorms could render AM reception virtually impossible across vast areas, while winter static from various electrical phenomena created a constant background of noise that severely compromised audio quality. Even in the absence of dramatic atmospheric disturbances, the accumulation of small interference sources—automobile ignition systems, electrical appliances, power lines, and countless other sources of electromagnetic interference—created a noise floor that limited the dynamic range and fidelity of AM broadcasting.

The static problem was particularly frustrating because it seemed to worsen precisely when radio was most popular. Evening hours, when families gathered around their receivers for entertainment, coincided with peak atmospheric activity that could make listening difficult or impossible. Summer months, when people had more leisure time for radio listening, were also the months when thunderstorm activity was at its highest. This seasonal and daily variation in reception quality made AM radio unreliable as a consistent source of high-quality entertainment.

Radio manufacturers had developed various techniques to mitigate the static problem, but none addressed its fundamental cause. Noise limiters could suppress the loudest static bursts, but they also distorted the audio signal and couldn't eliminate the constant background noise that plagued AM reception. Better antennas and more sensitive receivers could improve the signal-to-noise ratio by capturing stronger signals, but they couldn't eliminate the atmospheric interference that was inherent to amplitude modulation itself.

The audio quality limitations imposed by static had profound effects on AM programming and listener expectations. Music programming had to be carefully selected and processed to remain intelligible despite noise and interference. Complex orchestral arrangements often became muddy and indistinct when transmitted through AM systems, while quiet musical passages could be completely masked by background noise. This led to a preference for vocal music and simple instrumental arrangements that could penetrate the noise barrier, shaping the development of popular music in ways that reflected the technical limitations of the transmission medium as much as artistic preferences.

Armstrong's Frequency Revolution

Edwin Armstrong's approach to solving the static problem was characteristically innovative and theoretically elegant. Rather than trying to improve amplitude modulation by reducing its susceptibility to interference, Armstrong abandoned amplitude modulation entirely in favor of a completely different approach that encoded audio information in the frequency variations of the carrier wave rather than its amplitude variations.

In Armstrong's frequency modulation system, the carrier wave maintained constant amplitude while its frequency was varied in proportion to the audio signal being transmitted. When the audio signal increased in amplitude, the carrier frequency increased proportionally; when the audio signal decreased, the carrier frequency decreased correspondingly. This frequency deviation occurred around a center frequency, with the amount of deviation proportional to the amplitude of the modulating audio signal and the rate of deviation proportional to the frequency of the audio signal.

The genius of this approach lay in its immunity to amplitude-based interference. Since the audio information was encoded in frequency variations rather than amplitude changes, atmospheric disturbances that affected signal amplitude had no direct effect on the transmitted information. Lightning strikes, electrical interference, and other sources of amplitude-based noise could cause the received signal strength to fluctuate dramatically without affecting the frequency modulation that carried the audio content.

This fundamental difference in modulation technique required entirely new approaches to both transmission and reception. FM transmitters needed to generate stable carrier frequencies that could be varied precisely in response to audio input signals, requiring more sophisticated frequency control circuits than the relatively simple amplitude modulators used in AM systems. FM receivers needed to detect frequency variations rather than amplitude changes, requiring completely different demodulation circuits that could convert frequency changes back into audio signals.

The technical implementation of FM also revealed another crucial advantage: the possibility of much wider frequency deviation than was practical with AM systems. While AM broadcasting was limited to relatively narrow frequency bands to accommodate multiple stations in the available spectrum, FM could use much wider bands to achieve superior audio quality. This wider bandwidth made possible the transmission of the full audio frequency range with minimal distortion, enabling high-fidelity broadcasting that could reproduce music with unprecedented clarity and realism.

Armstrong's FM patents, granted in 1933, described not just a new modulation technique but an entirely new radio system from transmitter to receiver. These patents revealed the depth of Armstrong's understanding of the theoretical foundations of frequency modulation and demonstrated his ability to translate theoretical insights into practical engineering solutions. The patent documents showed that Armstrong had thought through all the implications of frequency modulation, from the mathematical relationships governing frequency deviation to the practical circuits needed to implement FM transmission and reception.

The Wide-Band Advantage

One of the most significant aspects of Armstrong's FM system was its use of wide-band frequency modulation, which provided superior performance compared to the narrow-band FM techniques that had been investigated by earlier researchers. Wide-band FM used much larger frequency deviations than narrow-band systems, typically varying the carrier frequency by 75 kilohertz or more in response to audio signals, compared to deviations of just a few kilohertz in narrow-band systems.

This wide frequency deviation provided several crucial advantages that made high-fidelity broadcasting practical for the first time. The wider deviation created a much more favorable signal-to-noise ratio, with the noise reduction proportional to the square of the deviation ratio. This meant that wide-band FM could provide dramatic improvements in audio quality compared to both AM broadcasting and narrow-band FM systems.

The wide bandwidth required by Armstrong's FM system initially seemed like a disadvantage, as it meant that fewer FM stations could operate in a given portion of the radio spectrum compared to AM stations. However, Armstrong recognized that this apparent disadvantage was actually a crucial element of FM's superior performance. The wide bandwidth was not wasted space but rather an essential component of the noise reduction mechanism that made high-fidelity broadcasting possible.

The frequency range available above 30 MHz provided ideal conditions for wide-band FM broadcasting. These Very High Frequency (VHF) bands offered sufficient bandwidth to accommodate multiple wide-band FM stations while providing propagation characteristics that were well-suited to high-quality broadcasting. VHF signals propagated primarily by line-of-sight transmission, eliminating the fading and interference problems associated with the sky-wave propagation that affected lower-frequency AM broadcasting.

The move to VHF frequencies also enabled FM broadcasting to escape the congested conditions that plagued the AM broadcast band. By 1933, the AM broadcast band was becoming increasingly crowded, with stations interfering with each other and creating reception problems that degraded audio quality. The VHF frequencies allocated to FM provided a clean spectrum environment where stations could operate without the interference problems that were becoming endemic in AM broadcasting.

The Industry's Resistance

Despite the obvious technical advantages of FM broadcasting, the established radio industry greeted Armstrong's invention with skepticism, resistance, and in some cases, outright hostility. This negative reaction reflected both legitimate technical concerns and economic interests that would be threatened by the widespread adoption of FM broadcasting.

The Radio Corporation of America (RCA), which had been Armstrong's primary partner in developing the superheterodyne circuit, initially showed interest in FM technology but quickly became one of its most determined opponents. RCA's resistance to FM reflected the company's enormous investment in AM broadcasting infrastructure and its commitment to developing television as the next major advancement in electronic communications. From RCA's perspective, FM represented a distraction from television development and a threat to the profitability of existing AM operations.

David Sarnoff, RCA's visionary leader who had previously supported Armstrong's superheterodyne development, famously dismissed FM as "a monumental waste of time" and refused to provide corporate support for its commercialization. This rejection forced Armstrong to pursue FM development independently, using his own financial resources and seeking support from smaller companies that were not committed to existing AM technologies.

The established broadcasters who had invested heavily in AM transmission equipment and programming also viewed FM with suspicion. Converting to FM would require completely new transmitters, new receivers for audiences, and potentially new programming approaches to take advantage of FM's superior audio quality. The investment required for this transition seemed enormous, especially during the economic uncertainties of the 1930s when many broadcasters were struggling financially.

Technical skeptics argued that FM's advantages were theoretical rather than practical, claiming that the superior audio quality promised by wide-band FM was unnecessary for typical radio programming. Many industry leaders believed that AM quality was adequate for speech and popular music, and that the additional fidelity offered by FM would not justify the costs of converting to the new system. This perspective failed to recognize FM's potential to transform radio into a high-fidelity medium that could compete with phonograph records for music reproduction quality.

The frequency allocation issues surrounding FM also created resistance from established interests. AM broadcasters argued that the VHF frequencies proposed for FM should be allocated to television instead, claiming that television represented a more important advance in communications technology. This argument reflected both genuine beliefs about television's importance and economic interests in protecting existing AM operations from FM competition.

Building the First FM Station

Faced with industry indifference and lacking corporate support, Armstrong made the remarkable decision to build and operate his own full-scale FM broadcasting station to demonstrate the practical advantages of his system. This decision represented an enormous personal and financial commitment that would ultimately cost Armstrong more than $300,000—equivalent to several million dollars in modern currency—and would consume years of his life in technical development and promotion activities.

The site Armstrong chose for his experimental FM station was Alpine, New Jersey, located on the Palisades across the Hudson River from New York City. This location provided excellent coverage of the New York metropolitan area while offering the elevation needed for effective VHF transmission. The station, initially designated W2XMN and later KE2XCC, began operations in 1939 with a 40-kilowatt transmitter—enormous power for the era—and a sophisticated antenna system designed to provide optimal coverage of the region.

The technical challenges of building the first FM station were formidable, as virtually every component had to be designed and built specifically for FM operation. No commercial FM transmitters were available, forcing Armstrong to design and construct his own transmission equipment. The modulation systems, frequency control circuits, and antenna matching networks all required innovative engineering solutions that pushed the boundaries of existing radio technology.

Armstrong's FM transmitter represented a masterpiece of radio engineering that incorporated numerous innovations in circuit design and construction techniques. The frequency stability required for high-quality FM transmission demanded precision oscillator circuits that exceeded anything previously used in broadcasting. The modulation systems had to provide linear frequency deviation across the entire audio frequency range while maintaining low distortion levels that would preserve the audio quality advantages of FM transmission.

The antenna system at Alpine was equally sophisticated, featuring a multi-bay array designed to provide uniform coverage of the service area while maximizing the effective radiated power. This antenna system established the pattern for future FM broadcasting installations and demonstrated the importance of proper antenna design for achieving optimal FM coverage.

Programming for the experimental FM station initially consisted of high-quality music broadcasts that showcased FM's superior audio reproduction capabilities. Armstrong recognized that demonstrating FM's advantages required programming that could reveal the differences between FM and AM broadcasting, leading to an emphasis on classical music, jazz, and other musical forms that could benefit from high-fidelity reproduction. These programs served both as technical demonstrations and as artistic showcases that revealed FM's potential as a medium for serious music broadcasting.

The Demonstration Campaign

Armstrong's experimental FM station became the centerpiece of an intensive campaign to demonstrate FM's advantages to the radio industry, government regulators, and the general public. This campaign involved countless demonstrations, technical presentations, and comparative tests designed to prove that FM could provide dramatically superior audio quality compared to AM broadcasting.

The demonstration techniques Armstrong developed became legendary in the radio industry for their dramatic effectiveness in revealing FM's advantages. One of the most famous demonstrations involved simultaneous broadcasts of the same program material on both AM and FM transmitters, with listeners switching between the two to hear the difference. The contrast was often startling, with the FM signal providing clear, noise-free reproduction while the AM signal suffered from static, interference, and limited frequency response.

Technical measurements supported the subjective impressions created by these demonstrations. Signal-to-noise ratio measurements showed that FM could provide noise levels 20 to 30 decibels lower than AM under typical reception conditions, representing a dramatic improvement in audio quality. Frequency response measurements revealed that FM could reproduce the full audio spectrum with minimal distortion, while AM systems suffered from significant frequency response limitations that affected music reproduction quality.

The demonstrations also revealed FM's superior performance under difficult reception conditions. During thunderstorms that rendered AM reception virtually impossible, FM continued to provide clear, noise-free signals that maintained full audio quality despite the atmospheric disturbances. This immunity to weather-related interference was particularly impressive to audiences who had learned to expect radio reception problems during storm conditions.

Professional musicians and audio engineers who heard these demonstrations were often among FM's most enthusiastic supporters, as they could immediately appreciate the difference in audio quality. Conductors, recording engineers, and classical music performers recognized that FM offered the possibility of radio broadcasting that could do justice to serious musical performances, potentially transforming radio from a medium for popular entertainment into a venue for high-quality cultural programming.

The technical press began covering Armstrong's FM demonstrations extensively, with detailed articles in engineering journals and trade publications documenting the superior performance of FM compared to AM. These technical reports provided scientific validation for the subjective impressions created by the listening demonstrations and helped establish FM's credibility within the engineering community.

Regulatory Battles and Frequency Allocations

The success of Armstrong's demonstrations led to regulatory proceedings that would determine FM's future in the American broadcasting system. The Federal Communications Commission (FCC), which had been established in 1934 to regulate all forms of electronic communication, found itself confronting the complex task of allocating frequency spectrum for FM broadcasting while balancing the competing interests of existing AM broadcasters, television developers, and FM proponents.

The initial regulatory hearings on FM broadcasting took place in the late 1930s and early 1940s, with Armstrong and other FM advocates presenting extensive technical evidence of FM's advantages. These hearings featured detailed technical testimony about FM's superior signal-to-noise ratio, immunity to atmospheric interference, and potential for high-fidelity broadcasting. The evidence presented was overwhelming in demonstrating FM's technical superiority for audio broadcasting applications.

However, the regulatory process was complicated by opposition from established broadcasting interests who argued that spectrum allocated to FM would be better used for television development. RCA and other major corporations presented arguments claiming that television represented a more important advance in communications technology and that FM broadcasting was an unnecessary luxury that would delay television development.

The compromise solution adopted by the FCC allocated spectrum in the 42-50 MHz band for FM broadcasting, providing sufficient bandwidth for a limited number of FM stations while preserving lower frequencies for AM broadcasting and higher frequencies for television. This allocation represented a victory for FM advocates, as it provided official recognition of FM as a legitimate broadcasting service with protected spectrum allocations.

However, the 42-50 MHz allocation proved to be temporary, as wartime developments and postwar policy changes would later force FM to relocate to higher frequencies. During World War II, the 42-50 MHz band was needed for military communications, and postwar frequency reallocations moved FM to the 88-108 MHz band where it operates today. This frequency change, while ultimately beneficial for FM's long-term development, required all existing FM stations to rebuild their transmission facilities and forced consumers to replace their FM receivers.

The frequency reallocation battle illustrated the political and economic forces that shaped broadcasting regulation during FM's early development. While technical considerations favored FM's adoption, regulatory decisions were influenced by powerful corporate interests that saw FM as a threat to their existing investments and future development plans. Armstrong found himself fighting not only technical battles but also political and legal battles that consumed enormous amounts of time, energy, and financial resources.

The War Years and Technical Development

World War II brought both opportunities and challenges for FM development. Military requirements for high-quality, interference-resistant communications led to extensive use of FM in military radio systems, validating Armstrong's claims about FM's technical advantages and spurring further development of FM technology.

The military applications of FM during World War II included tank communications, aircraft radio systems, and other applications where clear, reliable communication was essential for mission success. The immunity to atmospheric interference that made FM attractive for civilian broadcasting proved equally valuable for military communications, where interference could have life-or-death consequences.

Military FM development during the war years led to significant advances in FM transmitter and receiver technology. The need for portable, reliable equipment spurred innovations in circuit design, component miniaturization, and manufacturing techniques that would later benefit civilian FM broadcasting. Military research also advanced the understanding of VHF propagation characteristics and antenna design principles that were crucial for effective FM operation.

However, the war also delayed civilian FM development by diverting technical resources and manufacturing capacity to military production. FM broadcasting station construction was suspended for the duration of the war, preventing the expansion of FM service that might have accelerated public acceptance of the new medium. The wartime moratorium on civilian radio manufacturing also prevented the production of FM receivers needed to build the audience base for FM broadcasting.

Armstrong himself contributed to the war effort by working on military applications of FM technology and other radio developments needed for military operations. This work provided additional validation of FM's technical principles while generating income that helped support continued FM development. However, it also diverted Armstrong's attention from the commercial promotion of FM broadcasting at a crucial time in the technology's development.

The wartime experience with FM technology created a cadre of technically trained personnel who understood FM's advantages and would later support its civilian development. Radio technicians, engineers, and operators who had worked with military FM systems became advocates for civilian FM broadcasting when they returned to peacetime careers in the radio industry.

Post-War Renaissance and Growth

The end of World War II marked the beginning of FM's transition from experimental technology to commercial broadcasting medium. The post-war economic boom created favorable conditions for new technology adoption, while consumer demand for higher-quality entertainment led to increased interest in FM's superior audio quality.

The frequency reallocation that moved FM to the 88-108 MHz band, while initially disruptive, ultimately proved beneficial for FM's long-term development. The new frequency allocation provided more spectrum space for FM stations while offering better propagation characteristics and reduced interference from other services. The 88-108 MHz band became FM's permanent home and provided the foundation for the medium's eventual growth into a major broadcasting service.

The post-war period saw the beginning of serious commercial investment in FM broadcasting, as entrepreneurs and broadcasters began to recognize FM's potential for serving specialized audiences and providing programming that could differentiate itself from AM competition. Early FM programmers recognized that FM's superior audio quality made it particularly suitable for classical music, jazz, and other musical forms that could benefit from high-fidelity reproduction.

Educational and non-commercial broadcasters were among the early adopters of FM technology, as the lower costs of FM transmission equipment and the availability of reserved spectrum made FM an attractive option for institutions with limited budgets. Universities, foundations, and community organizations established FM stations that provided educational programming, classical music, and other content that was not well-served by commercial AM broadcasting.

The development of better FM receivers during the post-war period made FM more accessible to general audiences. Manufacturing improvements reduced receiver costs while improving performance, making FM receivers affordable for middle-class families. The introduction of combined AM/FM receivers allowed consumers to enjoy both broadcasting services with a single radio, reducing the barrier to FM adoption.

Stereo Broadcasting and High Fidelity

The ultimate vindication of Armstrong's vision for FM came with the development of stereophonic broadcasting, which transformed FM into the preferred medium for high-quality music reproduction. Stereo FM broadcasting, which began experimental transmission in the 1950s and became a commercial reality in the 1960s, provided audio quality that surpassed even high-quality phonograph records and established FM as the premier medium for serious music listening.

The technical foundation for stereo FM had been established by Armstrong himself, who had experimented with multiplexing techniques at his experimental station as early as 1934. These early experiments demonstrated that FM's wide bandwidth could accommodate not only high-quality monaural audio but also additional information channels that could carry supplementary services or, eventually, stereophonic audio channels.

The development of compatible stereo FM broadcasting required sophisticated engineering that maintained compatibility with existing monaural FM receivers while providing stereo capability for equipped receivers. The stereo multiplex system that was eventually adopted used a complex encoding scheme that transmitted left and right audio channels in a way that could be decoded by stereo receivers while remaining compatible with monaural reception.

Stereo FM broadcasting provided the audio quality that finally convinced music lovers and audiophiles that radio could be a serious medium for high-fidelity music reproduction. The combination of FM's inherent noise immunity with stereo reproduction created a listening experience that could rival the best home audio systems and provided programming quality that was impossible to achieve with AM broadcasting.

The success of stereo FM also validated the economic model that Armstrong had envisioned for FM broadcasting. By providing superior audio quality that justified premium advertising rates and attracted affluent audiences, stereo FM stations could generate the revenue needed to support high-quality programming and professional operations. This economic success encouraged further investment in FM broadcasting and accelerated the medium's growth throughout the 1960s and 1970s.

The Tragic End and Ultimate Triumph

Despite FM's eventual success, Edwin Armstrong would not live to see his invention reach its full potential. The patent litigation, industry opposition, and financial pressures that had dominated his later years took an enormous personal toll on the inventor who had revolutionized radio not once but twice during his career.

Armstrong's battle with RCA and other industry giants over FM patent rights became increasingly bitter and expensive during the post-war period. The legal fees and lost revenues from his FM patents consumed most of Armstrong's personal wealth, while the stress of prolonged litigation affected his health and personal relationships. The industry recognition and commercial success that he had earned from his earlier inventions seemed meaningless in the face of continued opposition to his greatest achievement.

In January 1954, facing mounting financial pressures and deteriorating health, Armstrong made the tragic decision to end his life. His suicide represented one of the great tragedies in the history of technology, as the man who had given the world high-fidelity broadcasting died just as his invention was beginning to achieve the success he had envisioned decades earlier.

Armstrong's death galvanized support for FM broadcasting and helped resolve the patent disputes that had consumed his final years. His widow, Marion Armstrong, continued the legal battles and eventually won significant patent settlements that validated Armstrong's contributions to FM development. More importantly, the technical community rallied around Armstrong's memory and worked to ensure that FM broadcasting received the recognition and support it deserved.

The years following Armstrong's death saw explosive growth in FM broadcasting, as stereo transmission, improved receivers, and changing musical tastes created the market conditions that Armstrong had predicted. By the 1970s, FM had surpassed AM in both audience size and advertiser support, becoming the dominant medium for music broadcasting and establishing itself as an essential component of the modern media landscape.

The Enduring Legacy

Edwin Armstrong's FM revolution represents one of the most significant technological achievements in the history of radio, transforming broadcasting from a medium limited by static and interference into a high-fidelity service capable of reproducing music with unprecedented clarity and realism. The technical principles that Armstrong established in his 1933 patents continue to govern FM broadcasting today, while his vision of radio as a medium for high-quality audio has been validated by decades of FM's dominance in music broadcasting.

The broader implications of Armstrong's work extend far beyond radio broadcasting itself. FM modulation techniques found applications in television sound transmission, mobile radio communications, satellite systems, and countless other technologies that require high-quality, interference-resistant signal transmission. The wide-band modulation principles pioneered by Armstrong became fundamental tools in the development of modern communication systems.

Perhaps most importantly, Armstrong's struggle to bring FM to the public demonstrated both the power of individual innovation and the institutional forces that can delay or distort technological progress. His story serves as a reminder that technical superiority alone is not sufficient to ensure the adoption of new technologies, and that economic, political, and social factors play crucial roles in determining which innovations succeed and which fail.

The clearer signals that Armstrong's FM revolution made possible transformed not only radio broadcasting but also the relationship between technology and culture. By enabling high-fidelity music reproduction through broadcasting, FM helped democratize access to high-quality audio and contributed to the development of the modern music industry. The FM revolution that began in Armstrong's laboratory ultimately changed how humanity experiences and shares music, creating possibilities for cultural expression and artistic communication that continue to enrich our lives today.

The invisible waves that carry FM signals continue to bring music, news, and entertainment to billions of listeners worldwide, fulfilling Armstrong's vision of radio as a medium worthy of the highest artistic and technical standards. In death as in life, Edwin Armstrong's greatest invention stands as a monument to the power of innovation to overcome both technical challenges and institutional resistance, providing clearer signals that connect humanity through the universal language of music and sound.

Chapter 9: The Amateur Radio Brotherhood

In the early hours of April 18, 1906, as the San Francisco earthquake shattered buildings and severed communication lines throughout the Bay Area, the regular channels of human communication—telephone lines, telegraph wires, and messenger services—lay in ruins beneath tons of rubble and twisted metal. Yet in those crucial first hours when disaster coordination meant the difference between life and death, an informal network of experimenters operating primitive wireless stations became the city's lifeline to the outside world. These amateur radio operators, working with spark-gap transmitters and crystal receivers built in basement workshops, provided the only reliable communication links that enabled rescue coordination and emergency response across the devastated region.

This dramatic demonstration of amateur radio's value during crisis marked the beginning of a tradition that would define the amateur radio community for more than a century: the willingness of radio amateurs to set aside their technical experiments and personal interests to serve their communities during emergencies. But the San Francisco earthquake also revealed something even more profound about amateur radio—it was creating a unique brotherhood of technically minded individuals who shared not only a fascination with electromagnetic waves but also a commitment to advancing the radio art for the benefit of all humanity.

The amateur radio community that emerged in the early twentieth century was unlike any other technical hobbyist group in history. United by their passion for wireless communication and their dedication to experimentation, radio amateurs developed a culture that emphasized technical excellence, mutual assistance, and public service. This brotherhood transcended national boundaries, political differences, and social distinctions, creating an international community of experimenters who freely shared their discoveries and collaborated on technical advances that would benefit not only amateur radio but the entire field of electronics and communications.

The Birth of a Movement

The origins of organized amateur radio can be traced to the experimental period following Heinrich Hertz's demonstration of electromagnetic waves and Guglielmo Marconi's development of practical wireless communication systems. As news of these wireless marvels spread through technical publications and popular magazines, thousands of experimenters around the world began building their own wireless apparatus and exploring the mysteries of radio wave propagation.

The first documented amateur radio station was established by Meade Dennis at Woolwich Arsenal in 1898, inspired by one of Marconi's lectures. Dennis's station represented the beginning of amateur radio as a distinct activity separate from commercial and military wireless operations. Unlike professional operators who used wireless for specific communication purposes, Dennis and other early amateurs were motivated primarily by technical curiosity and the thrill of wireless experimentation.

The amateur wireless movement gained momentum rapidly in the early 1900s as technical magazines published construction articles and circuit diagrams that enabled experimenters to build increasingly sophisticated wireless equipment. Publications like "Amateur Work," "Modern

Electrics," and "The Wireless Age" provided amateur operators with technical information, construction projects, and news about wireless developments around the world. These magazines also served as forums for amateur operators to share their experiences, compare results, and coordinate their experimental activities.

The formation of amateur radio clubs provided organizational structure for the growing amateur wireless community. The Wireless Telegraph Club of Columbia University, formed in 1908, was one of the earliest amateur radio organizations and established many of the traditions that would characterize amateur radio clubs for decades to come. These clubs provided meeting places for amateur operators, forums for technical discussions, and coordination centers for experimental activities and emergency communications.

The club movement spread rapidly throughout the United States and other countries as amateur operators recognized the benefits of organized cooperation. Clubs sponsored technical lectures, group construction projects, and competitive activities that encouraged technical excellence and innovation. They also served as training centers where experienced amateurs could share their knowledge with newcomers, ensuring the continued growth and technical advancement of the amateur radio community.

By 1912, the amateur radio movement had grown sufficiently large and organized to warrant formal government recognition and regulation. The Radio Act of 1912 established the first comprehensive regulatory framework for amateur radio in the United States, requiring amateur operators to obtain licenses and restricting their operations to specific frequency bands. While these regulations imposed some limitations on amateur activities, they also provided official recognition of amateur radio as a legitimate use of the electromagnetic spectrum and established the legal foundation for amateur radio's continued development.

The Great War Hiatus and Rebirth

World War I brought a dramatic interruption to amateur radio development that would profoundly influence the amateur community's character and priorities. In 1917, as the United States entered the war, all amateur radio operations were suspended and amateur radio equipment was sealed or confiscated by government authorities. This shutdown, which lasted until 1919, effectively ended amateur radio activity just as the movement was reaching maturity and beginning to demonstrate its potential for long-distance communication.

The wartime suspension of amateur radio had both immediate and long-term effects on the amateur community. Many amateur operators enlisted in military service, where their wireless expertise proved valuable for military communications. Others found civilian employment in the rapidly expanding wireless industry, applying their amateur radio experience to commercial and military wireless development. The amateur radio community that had been building for nearly two decades was scattered and its momentum disrupted.

However, the war also provided amateur operators with exposure to advanced wireless technologies that had been developed for military applications. Military wireless systems used higher power levels, more sophisticated circuits, and better understanding of propagation characteristics than had been available to amateur operators before the war. This exposure to advanced technology would prove invaluable when amateur operations resumed after the war, as returning amateur operators brought military wireless experience back to the amateur community.

The resumption of amateur radio operations in 1919 marked the beginning of a new era characterized by rapid technical advancement and international expansion. Amateur operators returning from military service brought new technical knowledge and a more systematic approach to wireless experimentation. The amateur radio community that emerged from the war was more technically sophisticated and more internationally minded than the pre-war amateur movement had been.

The post-war allocation of specific frequency bands to amateur radio—initially the 200-meter band and later shorter wavelengths—forced amateur operators to explore frequency ranges that professional radio engineers considered worthless for serious communication. This relegation to the "useless" short wavelengths would prove to be one of the most fortunate developments in amateur radio history, as amateur operators would soon discover that these frequencies possessed extraordinary capabilities for long-distance communication.

The Shortwave Revolution

The discovery by amateur radio operators in the early 1920s that shortwave frequencies could provide reliable long-distance communication represented one of the most significant technical breakthroughs in radio history. This discovery not only revolutionized amateur radio but also transformed international communication and established amateur operators as important contributors to the advancement of radio science.

Prior to 1920, amateur radio operators had been restricted to wavelengths of 200 meters and shorter (1,500 kHz and above), which professional radio engineers believed were unsuitable for long-distance communication. The conventional wisdom held that reliable long-distance radio communication required the massive power and enormous antennas associated with longwave transmission. Amateur operators, working with limited power and simple antennas, were expected to achieve only local communication within perhaps 25 to 50 miles.

But amateur operators approaching the wavelength restrictions with characteristic ingenuity began experimenting with increasingly shorter wavelengths, discovering that these frequencies could propagate far beyond the radio horizon under certain conditions. The first transatlantic amateur contact was achieved in 1921 when station 1BCG in Connecticut successfully communicated with station 8AB in France. This contact, achieved with modest power and simple equipment, demonstrated that shortwave frequencies could provide intercontinental communication capabilities that rivaled or exceeded the performance of the most powerful commercial longwave stations.

The implications of this discovery were staggering. Amateur operators had proven that the frequency bands they had been assigned were not worthless but were actually among the most valuable portions of the radio spectrum for long-distance communication. The shortwave bands that professional engineers had dismissed as useless could provide global communication with a tiny fraction of the power and antenna size required by conventional longwave systems.

The amateur radio community responded to this discovery with characteristic enthusiasm and systematic investigation. Amateur operators around the world began coordinating experiments to map shortwave propagation characteristics, identify optimal frequencies for different distances and times of day, and develop techniques for maximizing the effectiveness of shortwave communication. This informal but highly effective research program generated a vast amount of practical knowledge about shortwave propagation that would later be adopted by commercial and military radio services.

The success of amateur shortwave communication also demonstrated the value of the amateur radio community's experimental approach and international cooperation. Amateur operators freely shared their discoveries and collaborated across national boundaries to advance the understanding of shortwave propagation. This open exchange of information accelerated technical progress and established amateur radio as an important contributor to radio science.

International Organization and the IARU

The success of amateur shortwave communication made international coordination of amateur radio activities both necessary and practical. As amateur operators around the world established regular communication schedules and began participating in international experimental programs, the need for coordinated frequency allocations, operating procedures, and technical standards became apparent.

The International Amateur Radio Union (IARU) was formed in 1925 to provide international coordination for amateur radio activities and to represent amateur radio interests in international regulatory proceedings. The IARU brought together national amateur radio organizations from around the world and established the framework for international amateur radio cooperation that continues today.

The formation of the IARU represented a remarkable achievement in international cooperation, particularly considering the political tensions and communication difficulties of the 1920s. Amateur radio operators had created an international organization dedicated to technical cooperation and mutual assistance that transcended national boundaries and political differences. The IARU provided a model for international technical cooperation that would be emulated by other technical communities.

The IARU's early activities focused on frequency coordination and the development of international operating procedures that would minimize interference and maximize the effectiveness of amateur communications. The Union worked to establish band plans that allocated specific frequencies for different types of amateur activity, ensuring that experimenters working on similar projects could coordinate their activities and share their results effectively.

International amateur radio conventions and competitions sponsored by the IARU provided forums for amateur operators from different countries to meet personally, exchange technical information, and coordinate experimental activities. These gatherings fostered the international friendships and professional relationships that became one of amateur radio's most distinctive characteristics.

The IARU also served as the amateur radio community's representative in international regulatory proceedings, advocating for amateur radio frequency allocations and defending amateur radio against attempts to reallocate amateur frequencies to commercial or military services. This advocacy role required the IARU to demonstrate amateur radio's value to society and to document the technical contributions that amateur operators made to the advancement of radio science.

Technical Innovation and Experimentation

Throughout the 1920s and 1930s, amateur radio operators established themselves as important contributors to radio technology development through their systematic experimentation and willingness to explore unconventional approaches to technical problems. The amateur radio community's tradition of open information sharing and collaborative research created an environment that fostered innovation and accelerated technical progress.

Amateur radio operators made significant contributions to antenna design, developing directional arrays, vertical antennas, and other configurations that maximized performance while minimizing size and cost. The Yagi-Uda antenna, one of the most important directional antenna designs in radio history, was developed by Japanese amateur radio operators and shared freely with the international amateur community before being adopted for commercial and military applications.

Circuit development was another area where amateur operators made important contributions. Amateur experimenters developed improved receiver circuits, more efficient transmitter designs, and innovative approaches to frequency control and signal processing. The crystal-controlled oscillator, which became essential for frequency stability in all radio services, was perfected largely through amateur radio experimentation.

Modulation techniques were explored extensively by amateur operators, who experimented with various methods of encoding information onto radio waves. Amateur operators were among the first to develop practical single-sideband transmission techniques, which provided significant improvements in spectrum efficiency and power utilization compared to conventional amplitude modulation.

The development of very high frequency (VHF) and ultra high frequency (UHF) communication techniques was pioneered largely by amateur radio operators who explored these frequency ranges when they were considered impractical for serious communication applications. Amateur VHF experimentation revealed the line-of-sight propagation characteristics of these frequencies and demonstrated their value for local and regional communication networks.

Digital communication techniques were explored by amateur operators decades before they became commercially practical. Amateur radio pioneers developed packet radio systems, digital signal processing techniques, and computer-controlled station operation methods that anticipated many of the developments that would later revolutionize commercial communications.

The Sideband Revolution

One of amateur radio's most significant technical contributions was the development and perfection of single-sideband (SSB) transmission, a modulation technique that dramatically improved the efficiency of radio communication and enabled more effective use of the available radio spectrum. The development of practical SSB systems by amateur radio operators represented a major advance in communication technology that was later adopted by commercial and military radio services worldwide.

Traditional amplitude modulation, which had been used for voice communication since Fessenden's early experiments, transmitted the voice information in two identical sidebands on either side of the carrier frequency, along with the carrier itself. This approach was wasteful of both transmitter power and spectrum space, as two-thirds of the transmitted power was in the carrier (which carried no information) and half of the remaining power was in the redundant second sideband.

Amateur radio operators, working with limited power and increasingly crowded frequency bands, recognized that single-sideband transmission could provide significant advantages by eliminating the carrier and one sideband while retaining all the voice information in the remaining sideband. This approach could potentially provide the same communication effectiveness with one-fourth the transmitted power and half the bandwidth of conventional AM transmission.

The theoretical advantages of SSB transmission had been recognized by radio engineers since the early 1920s, but practical implementation required sophisticated circuit techniques that were beyond the capabilities of most amateur operators. The development of practical SSB systems required advances in frequency control, filtering techniques, and linear amplification that pushed the boundaries of available technology.

Amateur radio experimenters approached the SSB challenge with characteristic persistence and ingenuity, developing first the filter method and later the phasing method for generating SSB signals. These techniques required precision components and careful adjustment, but they enabled amateur operators to achieve SSB transmission with equipment that could be built in home workshops.

The advantages of SSB transmission became apparent immediately to amateur operators who adopted the technique. SSB signals could penetrate interference that would block conventional AM signals, while the improved power efficiency enabled effective communication with lower power levels. The reduced bandwidth requirements of SSB also meant that more stations could operate in the available frequency bands without interfering with each other.

The amateur radio community's development of practical SSB techniques attracted the attention of commercial and military radio engineers, who recognized the significant advantages that SSB could provide for their applications. Military adoption of SSB for tactical communications was particularly rapid, as the improved performance and spectrum efficiency were valuable for military operations.

Emergency Communications and Public Service

The tradition of public service that had emerged during the San Francisco earthquake became one of amateur radio's most distinctive characteristics and important contributions to society. Amateur radio operators around the world developed sophisticated emergency communication capabilities and established procedures for providing communication support during disasters and emergencies when normal communication systems failed or were unavailable.

The amateur radio emergency communication network that developed during the 1920s and 1930s was remarkable for its scope, effectiveness, and dedication. Amateur operators maintained emergency stations equipped with battery power supplies, emergency antennas, and portable equipment that could be deployed rapidly when disasters struck. Regional and national emergency networks provided coordination frameworks that could handle large-scale disasters requiring communication support across wide geographic areas.

The National Traffic System, developed by the American Radio Relay League (ARRL), provided a structured approach to handling emergency communications and routine message traffic through networks of amateur stations. This system enabled amateur operators to handle thousands of messages during major emergencies, providing crucial communication links between disaster areas and the outside world.

International emergency communication cooperation enabled amateur operators to provide communication support for disasters that crossed national boundaries or required international coordination. Amateur operators could establish communication links between disaster areas and international relief organizations, facilitating the coordination of international assistance and rescue operations.

The amateur radio community's emergency communication capabilities were tested repeatedly during natural disasters, transportation accidents, and other emergencies throughout the period. Each emergency provided lessons that led to improvements in equipment, procedures, and coordination techniques. The accumulated experience of the amateur emergency communication network created a body of practical knowledge about disaster communications that was unmatched by any other organization.

Training programs developed by amateur radio organizations ensured that amateur operators were prepared to provide effective emergency communications when needed. These programs emphasized not only technical proficiency but also message handling procedures, coordination techniques, and cooperation with served agencies such as the Red Cross and emergency management organizations.

The Brotherhood Culture

What distinguished amateur radio from other technical hobbies and professional activities was the unique culture of mutual assistance, technical sharing, and international friendship that developed among amateur operators. This culture, often referred to as the "amateur radio brotherhood," created bonds between operators that transcended national, racial, and social boundaries and established amateur radio as a force for international understanding and cooperation.

The amateur radio culture emphasized several core values that shaped the community's character and activities. Technical excellence was highly valued, with amateur operators striving to achieve the best possible performance from their equipment and to contribute to the advancement of radio technology. This emphasis on technical excellence drove continuous innovation and experimentation that benefited not only amateur radio but the entire field of electronics and communications.

Service to others was another fundamental value in amateur radio culture. Amateur operators were expected to provide assistance to other amateurs, to serve their communities during emergencies, and to contribute to the advancement of the radio art. This service orientation created a sense of responsibility and purpose that motivated amateur operators to contribute their time and expertise for the benefit of others.

International friendship and cooperation were emphasized as essential elements of amateur radio. Amateur operators were encouraged to make contacts with amateurs in other countries, to learn about different cultures and perspectives, and to build personal relationships that promoted international understanding. This international orientation was remarkable for its time and contributed to the development of a more internationally minded population.

The amateur radio community developed its own traditions, customs, and language that reinforced these cultural values. The use of call signs as personal identifiers created a sense of identity and belonging within the amateur community. Operating procedures and conventions emphasized courtesy, efficiency, and mutual respect. Technical discussions were conducted in a spirit of open inquiry and collaborative problem-solving.

Amateur radio clubs provided local focal points for the amateur community, offering meeting places, technical resources, and social activities that brought amateur operators together. These clubs sponsored technical presentations, group construction projects, and social events that strengthened the bonds within the amateur community and provided opportunities for experienced operators to mentor newcomers.

Elmers—experienced amateur operators who took responsibility for helping newcomers learn the technical and operating aspects of amateur radio—played a crucial role in perpetuating amateur radio culture and ensuring that new operators understood and embraced the community's values. The Elmer tradition created a mentoring system that was remarkably effective in transmitting both technical knowledge and cultural values from one generation of amateur operators to the next.

Regulatory Battles and Spectrum Defense

Throughout the period from 1930 to 1980, amateur radio operators faced numerous challenges to their frequency allocations as commercial and military services sought to expand their spectrum usage. The amateur radio community's response to these challenges demonstrated the effectiveness of organized advocacy and the importance of demonstrating amateur radio's value to society.

The most significant regulatory challenge faced by amateur radio during this period was the pressure to reallocate amateur frequencies to television broadcasting and other commercial services. As television broadcasting expanded in the post-World War II period, broadcasters argued that the VHF frequencies allocated to amateur radio could be better used for television stations serving major metropolitan areas.

Amateur radio organizations responded to these challenges by documenting amateur radio's contributions to radio technology development, emergency communications, and international goodwill. The amateur community's advocacy efforts emphasized amateur radio's role as a training ground for radio technicians and engineers, its contributions to emergency preparedness, and its value for promoting international understanding and cooperation.

The Geneva Radio Conference of 1959 marked a crucial victory for amateur radio, as international negotiations resulted in expanded frequency allocations for amateur radio and recognition of amateur radio's importance for technical development and emergency communications. These allocations provided amateur radio with secure spectrum access that enabled continued growth and development.

Regional radio conferences throughout the period required continued amateur radio advocacy to maintain existing allocations and secure additional spectrum for new amateur radio applications. Amateur radio organizations developed sophisticated technical presentations and political advocacy campaigns that effectively communicated amateur radio's value to regulatory authorities and the general public.

The amateur radio community's success in defending its spectrum allocations was due largely to the community's ability to demonstrate practical value through emergency communications, technical innovation, and international cooperation. This approach established amateur radio as a legitimate user of the radio spectrum and provided the foundation for amateur radio's continued development.

Technological Evolution and Modernization

The period from 1930 to 1980 witnessed dramatic technological changes that transformed amateur radio from a spark-gap and crystal detector hobby into a sophisticated electronic activity utilizing advanced solid-state technology and computer control. Amateur radio operators adapted to these technological changes while maintaining their tradition of experimentation and innovation.

The transition from vacuum tubes to transistors revolutionized amateur radio equipment design, enabling the development of portable and mobile equipment that would have been impossible with vacuum tube technology. Amateur radio manufacturers and experimenters led the development of transistorized amateur radio equipment, creating designs that were later adapted for commercial and military applications.

Integrated circuits further accelerated the modernization of amateur radio equipment, enabling the development of sophisticated transceivers with features that had previously been available only in professional equipment. Digital frequency displays, memory channels, computer control interfaces, and other advanced features became available to amateur operators as integrated circuit technology matured.

Microprocessor technology enabled amateur radio operators to experiment with computer-controlled station operation, digital signal processing, and automated communication systems decades before these techniques became commercially practical. Amateur radio's tradition of experimentation made it an ideal testing ground for new digital technologies and applications.

Satellite communication became an important new frontier for amateur radio experimentation as amateur operators launched their own satellites and developed ground station equipment for satellite communication. The amateur radio satellite program provided valuable experience with space communications that contributed to the development of commercial and military satellite systems.

Microwave communication techniques were pioneered largely by amateur radio operators working in the VHF, UHF, and microwave frequency bands. Amateur experimenters developed techniques for microwave propagation, antenna design, and system integration that were later adopted for commercial microwave communications systems.

The Legacy of Brotherhood

By 1980, amateur radio had evolved from a small community of wireless experimenters into an international movement encompassing hundreds of thousands of licensed operators around the world. The amateur radio brotherhood had demonstrated its value through technical contributions, emergency service, and international cooperation while maintaining its unique culture of mutual assistance and dedication to the advancement of the radio art.

The technical contributions of amateur radio operators during this fifty-year period were enormous and far-reaching. Amateur experimentation and innovation contributed to virtually every aspect of radio and electronics technology, from antenna design and circuit development to digital communications and microwave techniques. These contributions benefited not only amateur radio but also commercial communications, military systems, and the broader field of electronics engineering.

The emergency communication capabilities developed by amateur radio operators provided invaluable service during countless disasters and emergencies. The amateur radio emergency communication network had proven its effectiveness repeatedly and had earned recognition from government agencies and disaster relief organizations as an essential component of emergency preparedness.

The international cooperation fostered by amateur radio contributed to global understanding and friendship in ways that were difficult to quantify but impossible to ignore. Amateur radio contacts between operators in different countries created personal relationships that transcended political boundaries and contributed to international goodwill during periods of political tension.

The amateur radio brotherhood had also served as a training ground for countless engineers, technicians, and scientists who applied their amateur radio experience to professional careers in electronics and communications. Many of the most important developments in electronics and communications technology were created by people who had received their initial technical training through amateur radio.

As amateur radio entered the 1980s, the community faced new challenges from advancing technology, changing communication needs, and evolving spectrum requirements. But the amateur radio brotherhood that had been forged during the pioneering decades of radio development had proven its resilience and adaptability. The values and traditions that had guided amateur radio development for more than half a century would continue to inspire amateur operators as they explored new frontiers in digital communications, space technology, and the endless possibilities of the electromagnetic spectrum.

The invisible waves that connected amateur radio operators around the world carried more than just voice communications and digital data—they carried the spirit of innovation, service, and international friendship that defined the amateur radio brotherhood. This spirit would continue to guide amateur radio development and ensure that amateur radio remained a vital force for technical advancement and human understanding in the decades to come.

Chapter 10: Voices Across Borders - Propaganda and Politics

The static crackled across the shortwave bands on the evening of February 24, 1942, as technicians at the BBC's Broadcasting House in London made final adjustments to transmitters that would soon carry American voices across the Atlantic and around the world for the first time in history. In Studio B-4, a small room hastily converted for international broadcasting, a nervous announcer cleared his throat and prepared to speak words that would launch the Voice of America and forever change the nature of international communication. "Daily at this time," he began, "we shall speak to you about America and the war. The news may be good or bad. We shall tell you the truth."

With those simple words, the United States entered the international broadcasting arena that had been dominated by Nazi Germany's sophisticated propaganda machine and the BBC's more restrained information services. But the creation of VOA represented more than just America's belated recognition of radio's power as a tool of international influence—it marked the beginning of an invisible war fought not with bombs and bullets but with electromagnetic waves carrying competing visions of truth, ideology, and national purpose across the borders that had traditionally contained political discourse.

The transformation of radio from a neutral medium for entertainment and information into a weapon of political warfare revealed both the extraordinary power and the dangerous potential of mass communication technology. The same invisible waves that had brought families together around living room receivers in the 1920s would now be enlisted in a global struggle for hearts and minds that would span decades and reshape international relations in ways that the pioneering radio inventors could never have imagined.

The Early Seeds of International Broadcasting

The concept of using radio waves to influence foreign audiences had emerged gradually during the 1920s and 1930s as governments and organizations recognized that shortwave radio's ability to cross national boundaries made it a potentially powerful tool for international communication. Unlike newspapers, magazines, or films, which required physical distribution networks that could be controlled by national governments, radio signals respected no political boundaries and could reach audiences anywhere within their coverage areas.

The earliest international broadcasting efforts were motivated more by cultural exchange and information sharing than by deliberate propaganda objectives. The BBC's Empire Service, launched in 1932, was designed primarily to maintain connections between Britain and its far-flung colonial territories, providing news from home and reinforcing cultural ties across the British Empire. Similarly, the Netherlands' Radio Nederland and France's Radio Colonial were established to serve expatriate communities and colonial populations rather than to influence foreign public opinion.

However, the inherent nature of international broadcasting made purely neutral information transmission virtually impossible. The selection of news stories, the emphasis given to different events, and even the choice of language and cultural references inevitably reflected the perspectives and interests of the broadcasting nation. What began as innocent efforts to inform and entertain gradually evolved into more sophisticated attempts to shape international opinion and advance national interests through the careful manipulation of information and presentation.

The Nazi regime in Germany was among the first to recognize and systematically exploit radio's potential as a tool of international influence. German international broadcasting began modestly in the early 1930s but expanded rapidly as Hitler consolidated power and began pursuing aggressive foreign policies that required international justification and support. By 1939, Germany was transmitting nearly eleven hours of daily programming in multiple languages, with a significant portion of these broadcasts directed toward Britain and the United States.

The effectiveness of Nazi radio propaganda during the early years of World War II demonstrated both the power and the danger of international broadcasting. German broadcasters skillfully mixed accurate news reporting with subtle distortions and outright falsehoods, creating programming that was sufficiently credible to attract large audiences while gradually undermining confidence in democratic institutions and Allied war aims. The success of these broadcasts in neutral countries and even among Allied populations revealed that radio could be a formidable weapon in the battle for international opinion.

The Nazi Propaganda Machine

Joseph Goebbels, Hitler's propaganda minister, understood radio's potential for mass influence better than perhaps any other political leader of his era. Goebbels called radio the "eighth great power," recognizing that it could reach into every home and influence millions of people simultaneously in ways that no previous communication medium had ever achieved. Under his direction, Nazi Germany developed the most sophisticated and effective propaganda broadcasting system the world had yet seen.

The foundation of Nazi radio influence was the Volksempfänger, or "people's receiver," a cheap radio set subsidized by the government and distributed to German citizens to ensure maximum exposure to Nazi programming. These receivers were deliberately designed with limited shortwave capability to prevent German listeners from easily receiving foreign broadcasts while ensuring that they could receive German domestic programming clearly. The mass distribution of these receivers created a captive audience for Nazi propaganda while limiting exposure to alternative viewpoints.

Nazi international broadcasting strategy was remarkably sophisticated, employing psychological techniques and communication methods that would later be studied and emulated by intelligence agencies and propaganda organizations around the world. German broadcasters avoided the crude, obviously partisan approach that characterized much political communication of the era, instead developing programming that appeared objective and credible while subtly promoting Nazi viewpoints and undermining enemy morale.

The "black radio" operations conducted by Nazi Germany represented one of the most innovative and disturbing developments in propaganda broadcasting. These stations purported to be operated by resistance groups, dissident organizations, or independent journalists within enemy countries, but were actually controlled by German intelligence services and broadcast from German territory. By disguising their true origins, these stations could present anti-government messages that would be immediately dismissed if they were known to come from enemy sources.

German propaganda broadcasts to Allied countries employed sophisticated audience research and psychological profiling to maximize their effectiveness. Different programs were developed for different target audiences—British intellectuals received different messages than American factory workers, while broadcasts to colonial populations emphasized themes of liberation and racial equality that were designed to undermine support for European colonial powers. This targeted approach reflected a deep understanding of the diverse motivations and concerns that influenced different populations.

The notorious broadcasters who became known as "Lord Haw-Haw," "Tokyo Rose," and "Axis Sally" represented the personalization of enemy propaganda that made abstract political messages seem more intimate and credible. These English-speaking announcers, broadcasting from enemy territory, created an unsettling sense of intimacy with Allied audiences by demonstrating detailed knowledge of local conditions, military units, and even personal information about individual soldiers and civilians. Their broadcasts combined entertainment, seemingly credible news reporting, and subtle psychological manipulation in ways that were both effective and deeply disturbing to Allied authorities.

The Allied Response: VOA and the BBC

The success of Axis propaganda broadcasts created urgent pressure for the Allied powers to develop their own international broadcasting capabilities to counter enemy influence and present their own versions of events to global audiences. The British Broadcasting Corporation, with its established Empire Service and reputation for credible news reporting, expanded its international operations significantly during the war, while the United States launched the Voice of America as its first systematic effort at international broadcasting.

The creation of VOA in 1942 represented a fundamental shift in American foreign policy, marking the first time the U.S. government had attempted to systematically influence foreign public opinion through mass communication. The decision to enter the international broadcasting arena was driven by recognition that Nazi propaganda was succeeding in undermining Allied credibility and that radio had become an essential weapon in modern warfare. The Office of War Information took control of existing private shortwave operations and established VOA as America's official international radio voice.

VOA's founding philosophy, summarized in the promise to "tell the truth," reflected both idealistic American values and practical recognition that credibility was essential for effective international broadcasting. American planners understood that obviously biased propaganda would be quickly dismissed by sophisticated international audiences, and that long-term influence required building and maintaining audience trust through accurate and reliable reporting. This commitment to factual accuracy, while sometimes creating tensions with short-term political objectives, ultimately became one of VOA's greatest strengths.

The BBC World Service, evolved from the earlier Empire Service, took a somewhat different approach that emphasized cultural programming and comprehensive news coverage rather than explicit political messaging. The BBC's international broadcasts featured classical music, literary programs, educational content, and detailed news reporting that positioned Britain as a center of civilization and culture under attack by barbaric forces. This cultural emphasis proved particularly effective with educated audiences around the world who were attracted to British intellectual traditions and artistic achievements.

Both VOA and the BBC faced the challenge of balancing truthful reporting with wartime propaganda requirements. Military censorship prevented the broadcast of information that might aid enemy forces, while political considerations sometimes required emphasizing certain aspects of events while downplaying others. The skill with which Allied broadcasters navigated these constraints while maintaining their credibility became a crucial factor in the effectiveness of Allied international broadcasting.

The Armed Forces Radio Service (AFRS) represented another dimension of American radio's contribution to the war effort, providing entertainment and morale-boosting programming for American military personnel stationed around the world. While AFRS was not explicitly designed as a propaganda service, its high-quality entertainment programming and distinctly American cultural content helped project American values and lifestyle to international audiences who could receive its broadcasts. The irony that some of radio history's most star-studded programming was produced for military audiences and rarely heard by civilian listeners illustrated the extent to which radio had been mobilized for the war effort.

The Cold War Radio Battles

The end of World War II did not bring peace to the international broadcasting arena but rather transformed it into a new kind of battlefield where the Western democracies and the Soviet Union fought for global influence through competing radio services. The Cold War radio battles that emerged in the late 1940s represented a more sustained and systematic use of international broadcasting for political purposes than had ever been attempted during peacetime.

Radio Moscow, the Soviet Union's international broadcasting service, became VOA's primary adversary in a global competition for audience loyalty and ideological influence that would span four decades. Soviet international broadcasting had existed since the 1920s but expanded dramatically after World War II as Stalin's government recognized radio's potential for promoting communist ideology and undermining Western influence in developing countries and contested regions around the world.

The Soviet approach to international broadcasting differed significantly from Western models in both content and philosophy. While VOA and the BBC maintained at least nominal commitments to factual accuracy and balanced reporting, Radio Moscow explicitly embraced its role as a propaganda instrument and made no pretense of journalistic objectivity. Soviet broadcasts featured heavy-handed political commentary, conspiracy theories about Western imperialism, and selective reporting that consistently portrayed communist achievements favorably while emphasizing capitalist failures and contradictions.

Radio Free Europe and Radio Liberty represented a more aggressive Western approach to Cold War broadcasting that abandoned traditional journalistic models in favor of explicit political warfare objectives. These stations, secretly funded by the Central Intelligence Agency and other Western intelligence agencies, were designed to penetrate the Iron Curtain and provide alternative information sources for populations in communist countries who were denied access to independent media.

The clandestine nature of Radio Free Europe and Radio Liberty reflected the recognition that overt government broadcasting would be immediately dismissed by target audiences as foreign propaganda. By maintaining the fiction that they were operated by exile groups and independent organizations, these stations could present themselves as authentic voices of opposition rather than foreign intelligence operations. This deception was crucial to their effectiveness but also raised serious questions about the ethics of covert propaganda operations.

The technical aspects of Cold War radio warfare were as important as the content of the broadcasts themselves. Both sides invested heavily in high-power transmitters, sophisticated antenna systems, and advanced signal processing equipment designed to penetrate enemy jamming and reach target audiences despite deliberate interference. The electromagnetic spectrum became a literal battlefield where competing signals fought for dominance and audiences strained to hear forbidden voices through walls of electronic noise.

Jamming and Counter-Jamming

The Soviet Union's decision to jam Western radio broadcasts marked a new phase in the evolution of international broadcasting and demonstrated the extent to which radio had become a serious threat to authoritarian control. Jamming—the deliberate transmission of interfering signals designed to prevent reception of enemy broadcasts—represented a massive investment in electronic warfare that revealed both the power of radio propaganda and the vulnerability of closed societies to outside information.

Soviet jamming operations employed thousands of transmitters across the USSR and Eastern Europe, creating a vast network of interference sources that could blank out Western broadcasts across wide geographic areas. These jamming transmitters represented one of the largest electronic warfare installations ever constructed, consuming enormous amounts of electrical power and requiring constant maintenance by armies of technicians. The scale of the Soviet jamming effort provided clear evidence of the government's fear that foreign radio broadcasts posed a serious threat to political stability and ideological control.

The technical challenge of penetrating Soviet jamming drove continuous innovation in Western broadcasting technology and operations. Broadcasters developed frequency-hopping techniques that rapidly changed transmission frequencies to stay ahead of jamming signals, while high-power transmitters and sophisticated antenna systems were designed to overpower interference and deliver clear signals to target audiences. The technical arms race between broadcasters and jammers spurred advances in radio engineering that had applications far beyond the propaganda battlefield.

Western broadcasters also developed creative programming strategies designed to minimize the effectiveness of jamming while maximizing audience appeal. Short news bulletins repeated frequently on multiple frequencies made it difficult for jamming operations to completely suppress information flow, while entertainment programming mixed with political messages made broadcasts more attractive to audiences who might not listen to purely political content. The integration of popular music, cultural programming, and lifestyle information with political messages proved particularly effective in attracting young audiences in communist countries.

The psychological impact of jamming often backfired on the governments that employed it, as the obvious effort to suppress foreign broadcasts served as implicit confirmation that these broadcasts contained information that authorities feared their citizens might hear. The very existence of jamming signals reminded listeners that alternative information sources existed and encouraged them to seek out forbidden broadcasts despite the technical difficulties involved. This unintended consequence of jamming operations illustrated the complex dynamics of information warfare and the challenges faced by authoritarian governments in the radio age.

The Cultural Cold War

International broadcasting during the Cold War extended far beyond news and political commentary to encompass a broader cultural competition that used music, literature, educational programming, and lifestyle content to promote competing visions of modern civilization. This cultural dimension of radio warfare proved to be as important as explicitly political programming in shaping international opinion and influence.

American international broadcasting made extensive use of jazz, popular music, and entertainment programming to project images of American freedom, creativity, and cultural dynamism. The Voice of America's music programs, featuring leading American artists and performers, became popular with international audiences who were attracted to American cultural expressions even when they might be skeptical of American political messages. Jazz, in particular, became a powerful symbol of American creativity and individual expression that contrasted sharply with the regimented cultural expressions promoted by communist governments.

The BBC's cultural programming emphasized British intellectual traditions, classical music, and literary heritage to position Britain as a center of Western civilization and cultural achievement. BBC international broadcasts featured programming from major British cultural institutions—symphony orchestras, theatrical companies, universities, and literary societies—that reinforced Britain's role as a global cultural leader despite its declining political and military power. This cultural emphasis proved particularly effective with educated audiences around the world who respected British intellectual traditions.

Educational broadcasting became another important dimension of international radio competition as both sides recognized that knowledge and information were crucial weapons in the ideological struggle. Western broadcasters offered English language instruction, technical training, and educational programming that provided practical benefits to international audiences while subtly promoting Western values and perspectives. These educational services created loyal audiences who valued the practical benefits of Western broadcasts even when they might be skeptical of explicit political messages.

The Soviet Union and other communist countries responded with their own cultural and educational programming that emphasized socialist achievements in science, technology, and social organization. Radio Moscow featured programming about Soviet technological advances, educational achievements, and cultural productions that were designed to demonstrate the superiority of communist society. However, Soviet cultural programming often suffered from heavy-handed political messaging that reduced its appeal to international audiences who were primarily interested in entertainment and information rather than ideological instruction.

The Developing World Battleground

The emergence of newly independent nations in Africa, Asia, and Latin America during the 1950s and 1960s created new audiences and new battlegrounds for international radio competition. These developing countries, with limited domestic media infrastructure and populations hungry for information about the wider world, became prime targets for both Western and communist international broadcasting operations.

Both sides recognized that the political orientation of newly independent nations could significantly influence the global balance of power, making their populations crucial audiences for international broadcasting efforts. Western broadcasters emphasized themes of democratic development, free market economics, and individual liberty, while communist broadcasters promoted messages of anti-imperialism, socialist development, and solidarity with liberation movements. The competition for influence in the developing world gave international broadcasting a new urgency and strategic importance.

The technical challenges of reaching audiences in developing countries required significant investments in transmission facilities and specialized programming strategies. Many target countries lacked adequate electrical infrastructure, making battery-powered radios essential for mass communication. Broadcasters had to adapt their programming to accommodate limited receiver capabilities while competing with atmospheric interference and local stations for audience attention. The technical requirements of developing world broadcasting drove innovations in low-power transmission, efficient modulation techniques, and receiver design that had broader applications throughout the radio industry.

Language barriers created additional challenges for international broadcasters seeking to reach diverse audiences in countries with multiple ethnic groups and linguistic traditions. Major international broadcasting services expanded their language offerings dramatically during this period, with some services broadcasting in dozens of languages to reach specific ethnic and linguistic communities. This linguistic diversity required not only translation services but also cultural adaptation of programming content to reflect local customs, concerns, and interests.

The effectiveness of international broadcasting in developing countries was often enhanced by the limited availability of alternative information sources and the relatively weak domestic media infrastructure that characterized many newly independent nations. International radio broadcasts provided access to global news, cultural programming, and educational content that was simply unavailable through domestic sources. This information monopoly gave international broadcasters significant influence over public opinion and political development in many target countries.

From Information to Disinformation

The evolution of international broadcasting from relatively straightforward information services to sophisticated propaganda operations reflected broader changes in international relations and communication technology during the Cold War period. As both sides recognized the strategic importance of international opinion and the power of radio to influence it, international broadcasting became increasingly manipulative and deceptive in its methods and objectives.

The development of "gray" and "black" propaganda operations represented a significant departure from the relatively straightforward "white" propaganda that had characterized early international broadcasting. While white propaganda was openly identified with its sponsoring government and generally adhered to factual accuracy, gray and black propaganda operations used deception, misdirection, and outright falsehoods to achieve their objectives. These covert operations raised serious questions about the ethics and effectiveness of propaganda in democratic societies.

Radio Free Europe and Radio Liberty, despite their noble objectives of providing information to populations denied press freedom, employed techniques that would have been considered unethical in domestic journalism. These stations regularly broadcast unverified rumors, engaged in character assassination of communist leaders, and promoted political instability in target countries through carefully crafted disinformation campaigns. While these tactics may have been justified by the closed nature of communist societies, they established precedents for the use of international broadcasting as a tool of political warfare rather than information dissemination.

The psychological warfare aspects of Cold War broadcasting became increasingly sophisticated as both sides employed behavioral scientists, psychological researchers, and advertising professionals to maximize the effectiveness of their propaganda efforts. Market research techniques were adapted to measure audience preferences and responses, while psychological profiling helped broadcasters craft messages that would resonate with specific target populations. This scientific approach to propaganda made international broadcasting more effective but also more manipulative and potentially harmful to democratic discourse.

The blurring of lines between information and disinformation in international broadcasting had lasting effects on public trust in media institutions and international communication. Audiences became increasingly sophisticated in recognizing and discounting obvious propaganda, but the constant exposure to competing versions of truth and reality created cynicism and confusion that undermined confidence in all information sources. This erosion of shared factual foundations for public discourse created problems that would persist long after the end of the Cold War.

Technological Evolution and Global Reach

The technical infrastructure supporting international broadcasting expanded dramatically during the Cold War as both sides invested billions of dollars in transmission facilities, receiver development, and signal processing technology. High-power shortwave transmitters capable of generating millions of watts became common, while sophisticated antenna systems were designed to focus signals toward specific target areas with maximum effectiveness.

Satellite technology, which emerged in the 1960s, provided new opportunities for international broadcasting while also creating new challenges for governments seeking to control information flow. Satellite-delivered programming could reach areas that were difficult to serve with terrestrial shortwave transmitters, while also providing higher-quality signals that were less susceptible to jamming and atmospheric interference. However, satellite technology also made it more difficult for governments to control the reception of foreign programming, as satellite dishes could receive signals from multiple sources simultaneously.

The development of more sensitive and selective receivers made international broadcasting accessible to larger audiences while also enabling more sophisticated monitoring and analysis of foreign broadcasts. Transistor technology reduced the size and power consumption of shortwave receivers, making them more practical for use in remote areas and countries with limited electrical infrastructure. Digital signal processing techniques, which emerged in the 1970s, provided new capabilities for signal enhancement, interference suppression, and automatic tuning that made international broadcasting more user-friendly and effective.

Frequency management became increasingly complex as the number of international broadcasting stations multiplied and the available shortwave spectrum became crowded with competing signals. International coordination through the International Telecommunication Union helped minimize interference between different broadcasting services, but the demand for spectrum space continued to exceed supply throughout the Cold War period. This spectrum scarcity drove technical innovation in modulation techniques, antenna design, and signal processing that improved the efficiency of international broadcasting operations.

The measurement and analysis of international broadcasting effectiveness became a sophisticated science employing audience research techniques, signal strength measurements, and psychological assessment methods borrowed from commercial advertising and marketing research. Both sides invested heavily in monitoring facilities that could track the activities of competitor broadcasting services while measuring the effectiveness of their own operations. This research capability enabled more targeted and effective propaganda operations while also providing intelligence about competitor activities and capabilities.

The End of the Propaganda Era

The end of the Cold War in the late 1980s and early 1990s marked a dramatic transformation in international broadcasting that reflected broader changes in international relations and communication technology. The collapse of communist governments eliminated the primary rationale for much Cold War propaganda broadcasting, while new technologies like satellite television and the Internet created alternative channels for international communication that reduced radio's importance as a tool of international influence.

The reunification of Germany and the opening of Eastern Europe eliminated the need for many clandestine broadcasting operations while also providing opportunities for Western broadcasters to establish local partnerships and direct presence in formerly closed societies. Radio Free Europe and Radio Liberty adapted to these new conditions by establishing local offices and partnering with domestic broadcasters, transforming from clandestine propaganda operations into more conventional international news and information services.

The proliferation of independent media outlets in former communist countries reduced the importance of international broadcasting as a source of alternative information while also creating new competition for international broadcasters. Local radio stations, independent newspapers, and commercial television services could provide more relevant and timely information than distant international broadcasters, forcing international services to redefine their missions and target audiences.

The Voice of America and other Western international broadcasting services faced budget cuts and mission reviews as policymakers questioned the continued need for expensive international broadcasting operations in an era of reduced international tensions. Some services were eliminated or consolidated, while others were redirected toward new target audiences in regions where authoritarian governments continued to restrict press freedom and information access.

However, the skills and infrastructure developed during the Cold War radio battles found new applications in humanitarian broadcasting, development communication, and emergency information services. International broadcasters adapted their capabilities to serve populations affected by natural disasters, civil conflicts, and humanitarian crises, providing information and communication services that could save lives and coordinate relief efforts.

The Digital Transformation

The emergence of digital communication technologies in the 1990s and 2000s created new opportunities and challenges for international broadcasting while also fundamentally altering the relationship between governments, media institutions, and global audiences. The Internet, satellite television, mobile communications, and social media platforms provided new channels for international communication that were more difficult for governments to control and more responsive to audience preferences and needs.

International broadcasters adapted to these new technologies by developing websites, streaming services, and social media presences that complemented their traditional radio broadcasts while reaching new audiences who preferred digital communication channels. These digital platforms enabled more interactive and responsive communication with international audiences while also providing opportunities for user-generated content and community building that had been impossible with traditional broadcasting technologies.

The democratization of content creation and distribution enabled by digital technologies reduced the barriers to entry for international communication while also creating new challenges for established international broadcasting services. Independent journalists, citizen reporters, and activist organizations could reach international audiences directly without requiring expensive transmission facilities or government support, creating more diverse and competitive information environments in many target regions.

Social media platforms became new battlegrounds for international influence operations as governments and organizations recognized their potential for shaping public opinion and political outcomes. The techniques and strategies developed during the Cold War radio battles were adapted for use on Facebook, Twitter, and other platforms, creating new forms of information warfare that were more difficult to detect and counter than traditional broadcasting operations.

The proliferation of information sources and communication channels created new challenges for audiences seeking reliable and accurate information while also making traditional propaganda techniques less effective. The skills that audiences had developed for evaluating competing claims during the radio propaganda era proved valuable for navigating the more complex and chaotic information environments created by digital technologies.

The Enduring Legacy

The transformation of international radio from a neutral information medium to a weapon of political warfare during the twentieth century revealed both the extraordinary power and the dangerous potential of mass communication technology. The invisible waves that had initially promised to unite humanity in understanding and cooperation became tools for manipulation, deception, and ideological conflict that divided the world and undermined trust in information institutions.

Yet the international broadcasting infrastructure and professional expertise developed during the propaganda era also created valuable capabilities for humanitarian communication, emergency information services, and cross-cultural understanding that continue to serve important social purposes. The technical innovations spurred by international broadcasting competition contributed to advances in communication technology that benefit all users of the electromagnetic spectrum.

The ethical questions raised by Cold War propaganda broadcasting—about the responsibility of media institutions, the limits of acceptable persuasion, and the role of government in information dissemination—remain relevant in the digital age as new technologies create new opportunities for both communication and manipulation. The lessons learned from the radio propaganda era provide valuable insights for understanding and managing the information challenges of the twenty-first century.

Perhaps most importantly, the international broadcasting era demonstrated that information is indeed a form of power that can influence political outcomes, shape cultural development, and alter the course of human events. The voices that crossed borders through invisible radio waves carried more than news and entertainment—they carried competing visions of human society and different concepts of truth, freedom, and progress that would shape the modern world.

The radio waves that once carried the voices of propagandists and political warriors continue to circle the globe, now carrying messages of international cooperation, humanitarian assistance, and cultural exchange that fulfill radio's original promise of connecting humanity through the universal language of electromagnetic communication. The transformation from propaganda to partnership represents one of the most hopeful developments in the long history of humanity's relationship with the invisible waves that surround and connect us all.

Part IV: Beyond Earth – Radio Astronomy (1930-Present)

Chapter 11: Listening to the Universe

In the summer of 1931, as the Great Depression deepened and the world seemed to be coming unmoored from its foundations, a young radio engineer at Bell Telephone Laboratories was making a discovery that would fundamentally change humanity's relationship with the universe. Karl Jansky, just twenty-four years old and recently hired by Bell, was working on a routine task: identifying sources of static that might interfere with shortwave transatlantic voice transmissions. What he found instead was something that would open an entirely new "sight" for observing the universe—a discovery that would eventually reveal cosmic phenomena invisible to optical telescopes and transform our understanding of the universe itself.

The story of radio astronomy's birth is one of the most remarkable examples of scientific serendipity in history. Unlike many scientific breakthroughs that emerge from targeted research programs with specific objectives, radio astronomy was discovered almost by accident, by an engineer whose primary goal was to improve telephone service across the Atlantic. Yet this accidental discovery would prove to be one of the most significant developments in the history of astronomy, opening a new electromagnetic "window" on the universe that would reveal phenomena that had remained completely hidden from optical astronomers.

The Theoretical Foundation

The possibility of radio astronomy had existed in theory long before it became a practical reality. James Clerk Maxwell's equations of electromagnetism, published in the 1860s, had established that electromagnetic radiation could exist at any wavelength, from extremely long radio waves to extremely short gamma rays. Maxwell's mathematical framework showed that light was merely one small portion of a vast electromagnetic spectrum, and that other forms of electromagnetic radiation should exist beyond the narrow band visible to the human eye.

Despite this theoretical foundation, most physicists and astronomers in the late nineteenth and early twentieth century believed that astronomical radio waves would be undetectable from Earth's surface. The discovery of the ionosphere in 1902—the electrically charged layer of Earth's upper atmosphere—led scientists to conclude that any cosmic radio waves reaching Earth would be reflected back into space, making them impossible to detect from the ground. This theoretical barrier effectively prevented serious attempts to detect astronomical radio sources for several decades.

A few pioneering scientists did attempt to detect radio waves from the Sun in the late nineteenth and early twentieth century. In 1896, German astrophysicists Johannes Wilsing and Julius Scheiner conducted experiments to detect solar radio waves, and between 1897 and 1900, Oliver Lodge set up a centimeter wave radiation apparatus for similar purposes. However, these early attempts failed to detect any astronomical radio emission, primarily due to the technical limitations of the instruments available at the time. The sensitivity of early radio detectors was simply inadequate to detect the faint cosmic radio signals, and the theoretical understanding of radio wave propagation was still in its early stages.

The prevailing scientific opinion through the 1920s was that astronomical radio waves, even if they existed, would be too weak to detect with existing technology and would likely be reflected by the ionosphere before reaching Earth's surface. This widespread skepticism created a significant barrier to the development of radio astronomy, as few scientists were willing to invest time and resources in what seemed like a scientifically impossible endeavor.

Jansky's Unexpected Discovery

Karl Jansky's journey to the discovery of cosmic radio waves began with a practical engineering problem: improving the quality of transatlantic telephone service. In the late 1920s, Bell Telephone Laboratories had begun developing shortwave radio technology for transatlantic voice communications, which promised to be more reliable and higher quality than the existing telegraph-based systems. However, these shortwave transmissions were plagued by static interference that could render communications impossible during periods of high atmospheric activity.

In 1928, Bell assigned Jansky, a young engineer with expertise in radio technology, to investigate the sources of this static interference. Jansky's task was to identify and characterize the various sources of radio noise that might interfere with transatlantic communications, with the goal of developing methods to minimize or eliminate these interference sources.

To conduct his research, Jansky designed and built a large directional antenna system that could be rotated to determine the direction from which radio signals were coming. The antenna, which Jansky and his technicians called "Bromy" (a play on the word "Bromine" and the name of a Bell engineer), was a remarkable engineering achievement for its time. It consisted of a wooden framework covered with copper wire, supported by a set of four wheels from a surplus Navy surplus aircraft. The entire structure was rotated by a motor connected to a turntable, allowing Jansky to systematically survey the radio sky.

The antenna was connected to a sensitive radio receiver that recorded signal strength on a moving paper record. As Jansky conducted his systematic observations, he identified three distinct types of static: local thunderstorms, distant thunderstorms, and a faint, steady hiss of unknown origin. The first two sources were expected and relatively easy to identify, but the third source puzzled Jansky completely.

This mysterious "hiss" was much fainter than the thunderstorm static but remarkably persistent. Jansky initially suspected that the source might be the Sun, as the signal seemed to peak about every 24 hours. However, as he continued his observations, he noticed that the timing of the signal peaks was not precisely 24 hours but rather 23 hours and 56 minutes—the length of a sidereal day, which is the time it takes for the Earth to complete one rotation with respect to the "fixed" stars rather than the Sun.

This crucial detail led Jansky to realize that the source of the mysterious static was not the Sun but something beyond our solar system. By comparing his observations with optical astronomical maps, Jansky eventually determined that the radiation peaked when his antenna was aimed at the densest part of the Milky Way in the constellation of Sagittarius. In April 1933, Jansky published his results in the Proceedings of the Institute of Radio Engineers, reporting the discovery of "electromagnetic waves from outer space" emanating from the center of our Milky Way galaxy.

The First Cosmic Radio Waves

The significance of Jansky's discovery was profound, though it would take years for the scientific community to recognize its full implications. Jansky had detected radio waves with a wavelength of about 14.6 meters (corresponding to a frequency of approximately 20.5 MHz), which were being generated by some process in the center of our galaxy. This was the first time that astronomical phenomena had been observed using electromagnetic radiation outside the visible spectrum, opening a completely new field of astronomical investigation.

What Jansky had discovered was the radio emission from what is now known as Sagittarius A, the supermassive black hole at the center of our Milky Way galaxy. The radio waves he detected were generated by high-energy electrons spiraling through magnetic fields in the galactic center—a process now known as "solar" or "magnetospheric" radiation. This discovery revealed that the universe was filled with electromagnetic phenomena that were completely invisible to optical telescopes.

The technical challenges Jansky overcame to make this discovery were significant. His antenna system had to be sensitive enough to detect extremely faint signals while also being directional enough to determine their origin. The recording equipment had to be precise enough to distinguish between different types of static and to track the timing of signal peaks with sufficient accuracy to identify the sidereal day pattern. Jansky's background in both engineering and physics enabled him to design and build the necessary equipment while also understanding the astronomical implications of his observations.

Despite the significance of his discovery, Jansky faced skepticism from both the astronomical and radio engineering communities. Astronomers were accustomed to working with visible light and were skeptical of claims about radio waves from space, while radio engineers were primarily interested in practical communication applications rather than astronomical observations. Jansky's discovery was published in a radio engineering rather than an astronomical journal, limiting its initial impact on the astronomical community.

Reber's Radio Astronomy Revolution

While Jansky's discovery was groundbreaking, it might have remained a scientific curiosity if not for the work of Grote Reber, a young radio engineer who read about Jansky's discovery and became determined to follow up on it. Reber, working in his spare time in his backyard in Wheaton, Illinois, decided to build his own radio telescope to investigate the cosmic radio waves that Jansky had discovered.

Reber's efforts represented the first systematic attempt to study cosmic radio sources, and his work laid the foundation for the field of radio astronomy as a scientific discipline. Unlike Jansky, who had discovered cosmic radio waves while working on a practical engineering problem, Reber was deliberately pursuing astronomical research, making him the first true radio astronomer.

Reber's first challenge was to design and build a radio telescope that could detect cosmic radio waves more effectively than Jansky's antenna. After several attempts, he constructed a 9-meter (31.4-foot) parabolic reflector made of wooden strips covered with galvanized iron. This was the first parabolic radio telescope ever built and established the design that would become standard for radio astronomy for decades to come.

The technical challenges Reber faced were enormous. He had to design and build his own receiver, as commercial radio receivers of the time were not sensitive enough for astronomical observations. He also had to develop techniques for measuring and recording the extremely faint cosmic signals, which were often only a few percent stronger than the background noise.

Reber completed his radio telescope in September 1937 and began his observations. His first attempts to detect cosmic radio waves were not successful—he initially worked at 3300 MHz (9 cm wavelength), then at 900 MHz (33 cm), but found no detectable signal. It was only when he moved to a much lower frequency of 160 MHz (1.87 meters wavelength) in 1938 that he successfully detected cosmic radio waves, confirming Jansky's discovery.

Over the next several years, Reber conducted systematic observations of the radio sky, creating the first radio frequency sky map. His observations revealed that the radio emission from the Milky Way was not uniform but concentrated in specific regions, with the most powerful source coming from the galactic center. Reber also discovered radio emission from other parts of the sky, suggesting that cosmic radio sources were more widespread than initially thought.

In 1940, Reber published his first professional paper in the Astrophysical Journal, bringing radio astronomy to the attention of the broader astronomical community. His work demonstrated that radio astronomy was not just a curiosity but a legitimate field of astronomical research with the potential to reveal new information about the universe. Despite being offered a position at Yerkes Observatory, Reber chose to continue his independent research, further developing the techniques and methodology of radio astronomy.

The War Years and Post-War Revolution

World War II temporarily disrupted the development of radio astronomy as scientific resources were redirected toward military applications. However, the war also accelerated the development of radio technology in ways that would ultimately benefit radio astronomy. The development of radar systems required advances in radio receiver technology, antenna design, and signal processing that would later prove invaluable for radio astronomy.

During the war, British radar operators made an important discovery that would have significant implications for radio astronomy. In 1942, British radar stations detected mysterious radio interference that was initially thought to be German jamming. However, further investigation revealed that the interference was coming from the Sun, specifically during periods of high solar activity. This discovery of solar radio emission was kept secret during the war but was published after the conflict ended, revealing that the Sun was a powerful source of radio waves.

The end of World War II marked the beginning of a revolution in radio astronomy as scientists who had worked on radar and other radio technologies turned their attention to astronomical applications. The availability of surplus radar equipment, combined with the technical expertise developed during the war, created ideal conditions for the rapid expansion of radio astronomy.

In Australia, a group of scientists led by J. G. "Taffy" Loveband used surplus radar equipment to establish one of the first dedicated radio astronomy programs. Their observations revealed that the Sun was not the only astronomical source of radio waves—distant cosmic sources were also producing strong radio emission. These discoveries quickly attracted the attention of astronomers around the world, leading to the establishment of radio astronomy programs in multiple countries.

In Britain, scientists at Cambridge University, including Martin Ryle and his group, began developing interferometric techniques that dramatically improved the directional accuracy of radio telescopes. These techniques, which combined signals from multiple radio telescopes to create a single high-resolution instrument, would become fundamental to radio astronomy and eventually lead to the development of radio interferometric arrays that could achieve unprecedented levels of detail in radio observations.

The Discovery of Radio "Sirens" in the Sky

One of the most significant early discoveries in radio astronomy was the detection of extremely powerful radio sources that appeared as points of light in radio maps but had no visible optical counterpoints. These mysterious objects, initially called "Radio Sirens" because of their powerful radio emission, would eventually be identified as some of the most exotic and powerful objects in the universe.

The first of these objects, designated as "C" in the early Cambridge radio source list, was detected in 1948 but not properly identified until 1951. When optical astronomers finally located the source, they discovered it was associated with a faint, star-like object that showed an enormous shift in its spectrum toward longer wavelengths—a "red shift" that suggested it was moving away from us at an extraordinary speed. This object, later named 3C 48, was the first identified "quasar" (quasistellar radio source), a type of object that would revolutionize our understanding of the universe.

The discovery of additional radio sources with no obvious optical counterpoints led to a major effort to identify these mysterious objects. By the mid-1950s, radio astronomers had created detailed radio source maps, and optical astronomers were working to identify the optical sources associated with these radio objects. This collaborative effort between radio and optical astronomers revealed that many of these radio sources were associated with distant, active galactic nuclei—massive black holes at the centers of distant galaxies that were consuming enormous amounts of matter and producing powerful radio emission in the process.

The discovery of these powerful radio sources demonstrated that radio astronomy was not just a complementary tool for optical astronomy but a means of discovering entirely new types of astronomical objects that were invisible to optical telescopes. The universe, as revealed by radio waves, was far more dynamic and exotic than astronomers had previously imagined.

The Cosmic Microwave Background

Perhaps the most profound discovery in radio astronomy came completely by accident in 1964, when two Bell Telephone Laboratories scientists, working on satellite communication technology, stumbled upon evidence of the Big Bang itself. This discovery would provide the most direct evidence for the theory that the universe began in a massive explosion and has been expanding ever since.

In 1964, radio astronomers and engineers at Bell Telephone Laboratories were working on the development of satellite communication systems using microwave technology. Among them were two scientists, Arno A. Penzias and R. W. "Ric" Wilson, who were attempting to improve the sensitivity of a large microwave antenna originally built for satellite communication experiments.

Penzias and Wilson were using the antenna to study radio waves from the Milky Way, but they kept detecting a persistent background noise that seemed to come from all parts of the sky. This noise was extremely faint—equivalent to the microwave radiation that would be produced by an object with a temperature of about 3.5 K (3.5 degrees above absolute zero)—and it appeared to be completely uniform in all directions.

The two scientists initially suspected that the noise was caused by some problem with their equipment. They carefully eliminated all possible sources of interference, including the famous "pigeon problem"—a pair of white-footed doves that had built a nest in the antenna and left behind a layer of "white, gray, and black material" that they had to clean out. Despite their efforts, the mysterious background noise remained.

Penzias and Wilson eventually learned that scientists at nearby Princeton University, led by physicist R. H. Dicke, were actually looking for exactly this kind of background radiation as evidence for the Big Bang theory of the universe's origin. Dicke and his group had predicted that if the universe began in a massive explosion, there should be a faint background of microwave radiation left over from that event, cooled by the expansion of the universe to a temperature of just a few degrees above absolute zero.

When Penzias and Wilson compared their measurements with Dicke's predictions, they realized they had discovered the cosmic microwave background (CMB) radiation—the afterglow of the Big Bang itself. This discovery provided the most direct evidence for the Big Bang theory and revolutionized our understanding of the universe's origin and evolution.

Penzias and Wilson published their discovery in 1965, and in 1978 they were awarded the Nobel Prize in Physics for their work. The discovery of the CMB radiation remains one of the most important in the history of astronomy, providing a "fossil" record of the early universe that has allowed scientists to study conditions just 380,000 years after the Big Bang.

The Evolution of Radio Astronomy Technology

The development of radio astronomy required significant advances in technology, as traditional optical telescope designs were not suitable for observing radio waves. Radio waves have much longer wavelengths than visible light—ranging from millimeters to meters rather than fractions of a micrometer—which means that radio telescopes need to be much larger to achieve the same level of detail in observations.

The basic design of a radio telescope is relatively simple: a large reflecting surface (often parabolic) that captures radio waves and reflects them to a focal point where a receiver is located. However, building large, precise reflecting dishes presented significant engineering challenges, especially as astronomers sought to build larger telescopes to detect fainter radio sources.

One of the most significant technological advances in radio astronomy was the development of interferometric techniques, which allow multiple smaller radio telescopes to work together as a single, much larger telescope. By carefully measuring the time delay between when a radio wave from a cosmic source reached each telescope in an array, astronomers could effectively create a telescope with a size determined by the maximum distance between the individual telescopes.

The Very Long Baseline Interferometric (VLSI) technique, developed in the 1960s, took this concept to its extreme by using radio telescopes separated by thousands of kilometers as a single instrument. This technique, which required extremely precise atomic timekeeping to coordinate observations across vast distances, allowed radio astronomers to achieve unprecedented levels of detail in their observations, effectively creating a radio telescope the size of the Earth.

The development of space-based radio astronomy represented another major technological advance. While Earth's atmosphere is relatively clear to radio waves (unlike many other wavelengths of electromagnetic radiation), certain frequencies are still affected by the ionosphere or by interference from human activities. Space-based radio telescopes, such as the Soviet Union's Relikt-1 satellite launched in 1983 to study the cosmic microwave background radiation at 37 GHz, allowed astronomers to observe at frequencies that were difficult or impossible to observe from the ground.

The launch of NASA's Cosmic Background Explorer (COBE) satellite in 1989 represented a major advance in the study of the cosmic microwave background. COBE was able to measure the spectrum of the CMB with unprecedented precision, confirming that it matched the theoretical prediction for a black-body radiation spectrum with a temperature of 2.73 K. This confirmation provided strong evidence that the CMB was indeed the remnant radiation from the Big Bang, rather than some other phenomenon.

The Radio Universe Revealed

As radio astronomy technology advanced, it revealed a universe far more dynamic and exotic than optical astronomers had previously imagined. Radio observations showed that the universe was filled with high-energy phenomena that produced powerful radio emission through mechanisms that were often completely different from those producing visible light.

One of the most significant discoveries was that of "pulsars"—highly magnetized, rotating neutron stars that produce regular pulses of radio waves. Discovered in 1967 by Jocelyn Bell and Martin Ryle at Cambridge University, these objects were initially so mysterious that they were given the temporary name "LGM" (for "Little Green Men") because their regular pulses seemed too precise to be natural. It was soon determined that these pulses were produced by rapidly rotating neutron stars, the extremely dense remains of massive stars that had exploded as supernovae.

Radio observations also revealed the existence of "quasars"—quasistellar radio sources that were later determined to be the extremely active nuclei of distant galaxies. These objects, powered by supermassive black holes consuming enormous amounts of matter, were among the most powerful energy sources in the universe, producing more energy than entire galaxies while being no larger than our solar system.

The discovery of "masing" in space—astronomical analogs to lasers that produce powerful microwave emission—was another significant radio astronomy discovery. These natural microwave amplification by the stimulated emission of radiation (or "masing") phenomena occur in regions of star formation and around evolved stars, providing astronomers with powerful tools for measuring distances and understanding the physics of these regions.

Radio observations of neutral atomic hydrogen (the most common element in the universe) using the 21 cm hydrogen line have allowed astronomers to map the structure of our Milky Way galaxy and other galaxies in unprecedented detail. These observations revealed that galaxies are often embedded in vast, rotating hydrogen gas disks that extend far beyond the visible stars, providing crucial evidence for the existence of "dark matter" that makes up most of the mass of galaxies.

The Impact on Modern Astronomy

The impact of radio astronomy on modern astronomy cannot be overemphasized. Radio observations have provided crucial insights into virtually every area of astronomical research, from the study of our solar system to the large-scale structure of the universe.

In our solar system, radio observations have been used to study the atmosphere and surface of other planets, to detect water and other complex molecules in interstellar space, and to search for signs of extraterrestrial intelligence through the SETI (Search for Extraterrestrial Intelligence) program. Radio telescopes have even been used to bounce signals off other planets and measure their precise distances, providing crucial data for testing theories of general relativity.

Radio astronomy has been particularly important for the study of the early universe. The cosmic microwave background radiation, discovered through radio observations, has provided astronomers with a detailed "fossil record" of the universe when it was just 380,000 years old. More recently, radio observations of the "dark age" of the universe—the period before the first stars formed—through the detection of the 21 cm hydrogen line have begun to reveal the conditions in the universe during its first few million years.

The development of radio astronomy has also had significant practical applications beyond pure scientific research. The technology developed for radio astronomy has been applied to satellite communication, radar technology, and even the development of the World Wide Web (which was originally created at CERN to share data among high-energy physics researchers). The precise timing technology developed for very long baseline interferometric observations has been applied to global satellite-based position systems like the Global Positioning System (GPS).

The Future of Radio Astronomy

As radio astronomy continues to evolve, it is opening new frontiers in our understanding of the universe. The development of next-generation radio telescopes, such as the Square Kilometre Array (which will have a total receiving area of one square kilometer), will provide unprecedented sensitivity and detail in radio observations. These new instruments will allow astronomers to study the formation of the first stars and galaxies, to search for signs of life on other planets, and to test fundamental theories of physics under extreme conditions.

One of the most ambitious projects in radio astronomy is the search for gravitational waves through the timing of millisecond pulsars. By carefully measuring the arrival times of pulses from an array of rapidly rotating neutron stars, astronomers hope to detect the subtle distortions in space caused by passing gravitational waves from massive cosmic events such as the merger of supermassive black holes.

Radio astronomy is also playing a crucial role in the search for extraterrestrial intelligence. The SETI program, which began in 1960 with the first systematic search for radio signals from other technological societies, continues to search for evidence of life beyond Earth using increasingly sophisticated radio telescopes and signal processing techniques.

The development of new data processing techniques, including machine learning and artificial intelligence, is allowing radio astronomers to extract more information from their observations than ever before. These techniques are helping astronomers to identify rare and subtle phenomena in the vast amounts of data produced by modern radio telescopes, potentially leading to new discoveries that would have been impossible with traditional analysis methods.

The Symphony of the Universe

As we reflect on the history of radio astronomy, from Jansky's accidental discovery to the sophisticated global networks of radio telescopes that now study the universe, we are reminded of the profound truth that the universe is not just a place we see—it is also a place we can hear. The invisible waves that carry our radio broadcasts and television signals are the same waves that carry the "symphony of the universe"—the cosmic radio emission that reveals the dynamic, high-energy phenomena that shape the evolution of galaxies, stars, and planets.

Radio astronomy has given humanity a new sense with which to experience the universe—a sense that reveals phenomena invisible to optical telescopes and allows us to "eavesdrop" on cosmic events that occurred billions of years in the distant past. Through radio waves, we have heard the echoes of the Big Bang itself, the regular pulses of neutron stars, the powerful radio emission from supermassive black holes, and the faint signals from the earliest stars and galaxies.

The story of radio astronomy is a powerful reminder that scientific discovery often comes from unexpected places and that the pursuit of knowledge for its own sake can lead to profound insights about our place in the universe. Jansky was not looking for cosmic radio waves when he began his work at Bell Telephone Laboratories—he was simply trying to improve telephone service. Yet his discovery opened a new "window" on the universe that has transformed our understanding of cosmic phenomena.

As we continue to develop more sensitive radio telescopes and more sophisticated analysis techniques, we can only imagine what new cosmic "symphony" we might hear in the future. The universe is filled with invisible waves carrying information about its most dramatic events and fundamental properties, and radio astronomy has given us the ability to listen. In doing so, it has not only expanded our scientific knowledge but has also deepened our sense of connection to the vast, dynamic universe of which we are a part.

Chapter 12: Breakthrough Discoveries from Space

The Cambridge sky was overcast and dreary on that November evening in 1967 when graduate student Jocelyn Bell sat hunched over rolls of chart paper in a cramped office, her eyes straining to examine the continuous traces of radio signals recorded by the university's new radio telescope array. For months, she had been methodically analyzing these paper records, searching for the faint radio signatures of distant quasars that flickered due to interstellar turbulence—a phenomenon called scintillation. What she discovered instead would revolutionize our understanding of stellar death and the extreme states of matter possible in the universe.

Among the countless squiggles and variations on the chart paper, Bell noticed something extraordinary: a series of perfectly regular pulses occurring with clockwork precision every 1.3373 seconds. The signals were so regular, so artificial-seeming, that Bell and her supervisor Antony Hewish initially wondered if they had detected signals from an extraterrestrial civilization. They even gave the source the whimsical designation "LGM-1"—short for "Little Green Men". But as Bell continued her painstaking analysis and discovered three more similar sources scattered across the sky, it became clear that these mysterious pulsing radio beacons represented not alien technology but something far more profound: the discovery of an entirely new class of astronomical objects that would open unprecedented windows into the physics of matter under the most extreme conditions imaginable.

The discovery of pulsars—as these objects came to be known—marked the beginning of a golden age of radio astronomy discoveries that would fundamentally transform our understanding of the universe. Over the following decades, radio telescopes would reveal a cosmos far more dynamic, violent, and exotic than anyone had previously imagined, populated by supermassive black holes, colliding galaxies, and phenomena that pushed the very laws of physics to their limits.

The Pulsar Revolution

The realization that Bell's mysterious radio pulses represented a new type of astronomical object came gradually, as astronomers around the world struggled to understand what could produce such incredibly regular signals from space. The precision of the pulses was extraordinary—more regular than the best atomic clocks of the era—yet they originated from points of light so faint that they were barely detectable with optical telescopes.

The breakthrough in understanding pulsars came when astronomers realized that these objects were rapidly rotating neutron stars—the collapsed cores of massive stars that had exploded as supernovae. This connection had been predicted theoretically decades earlier by Walter Baade and Fritz Zwicky in 1934, who had proposed that supernovae represented the collapse of ordinary stars into incredibly dense neutron star structures. However, neutron stars had remained purely theoretical objects until Bell's discovery provided the first observational evidence of their existence.

The physics of neutron stars stretched the imagination. These objects contained more mass than our Sun compressed into a sphere only about 20 kilometers in diameter—so dense that a teaspoon of neutron star material would weigh about a billion tons on Earth. The extreme density created gravitational fields so intense that the escape velocity from the surface approached 100,000 kilometers per second—about one-third the speed of light. The magnetic fields surrounding these objects were trillions of times stronger than Earth's magnetic field, powerful enough to distort the very atoms in their vicinity.

The pulsar mechanism itself was elegantly simple yet profoundly revealing about the nature of these exotic objects. As neutron stars rotate, their powerful magnetic fields accelerate charged particles to nearly the speed of light, creating beams of radio emission that sweep across space like cosmic lighthouses. When these beams happen to point toward Earth, radio telescopes detect the characteristic pulses that had captured Bell's attention. The incredible regularity of these pulses reflected the precise rotation of objects whose moment of inertia was so enormous that they could maintain constant rotation rates for millions of years.

The discovery of the Crab Pulsar in 1968 provided dramatic confirmation of the neutron star theory. Located at the center of the Crab Nebula—the remnant of a supernova explosion observed by Chinese astronomers in 1054 AD—this pulsar was spinning 30 times per second and gradually slowing down as it radiated energy into space. The connection between this pulsar and the historical supernova provided compelling evidence that pulsars were indeed the neutron star remnants of stellar explosions, exactly as Baade and Zwicky had predicted three decades earlier.

The study of pulsars opened entirely new fields of astrophysics, allowing scientists to investigate fundamental physics under conditions impossible to replicate on Earth. Pulsar timing became one of the most precise measurement techniques in astronomy, enabling tests of Einstein's general relativity theory and the detection of gravitational waves through their effects on pulsar signals. The discovery of binary pulsars—neutron stars orbiting other compact objects—provided laboratories for studying the behavior of matter and energy under the most extreme gravitational conditions in the universe.

The Quasar Enigma

While Bell was analyzing pulsar signals in Cambridge, astronomers around the world were grappling with another radio astronomy mystery that would prove equally revolutionary: the nature of extremely powerful radio sources that appeared as star-like points of light but exhibited properties unlike any known astronomical objects. These "quasi-stellar radio sources," or quasars, would challenge fundamental assumptions about the scale and energy production mechanisms in the universe.

The quasar story began in the late 1950s when radio astronomers started creating detailed catalogs of radio sources across the sky. Among these sources were several that appeared to be associated with faint, star-like optical objects that showed no obvious characteristics distinguishing them from ordinary stars. However, when astronomers obtained detailed spectra of these objects, they discovered something extraordinary: the spectral lines were shifted toward longer wavelengths by enormous amounts, indicating that these objects were moving away from us at significant fractions of the speed of light.

The first identified quasar, 3C 273, showed a redshift indicating it was receding at about 16% of the speed of light. If this redshift was due to the expansion of the universe—as it appeared to be—then 3C 273 was located at a distance of over two billion light-years from Earth. At such an enormous distance, the object would have to be producing more energy than an entire galaxy of stars just to appear as bright as it did in radio observations.

This realization created what became known as the "energy crisis" in astronomy. No known physical process could account for the incredible energy output that quasars would require if they were indeed at cosmological distances. The most luminous quasars appeared to be producing energy equivalent to the combined output of hundreds of billions of stars, yet they varied in brightness over periods of months or even weeks—implying that the energy-producing region was no larger than our solar system.

The resolution of the quasar energy crisis came with the recognition that these objects were powered by supermassive black holes—gravitational monsters containing millions or billions of times the mass of our Sun. As matter spiraled into these black holes, it would be heated to millions of degrees and accelerated to enormous speeds, producing intense radiation across the electromagnetic spectrum. The efficiency of this accretion process meant that a relatively small amount of matter falling into a black hole could produce the enormous energy outputs observed in quasars.

The discovery of quasars revealed that the early universe was a far more violent and energetic place than astronomers had previously imagined. Most quasars were located at great distances, meaning that we were observing them as they existed billions of years ago when the universe was much younger. This suggested that supermassive black holes and the violent processes associated with them were common in the early universe, playing crucial roles in the formation and evolution of galaxies.

The study of quasars also provided new tools for investigating the large-scale structure of the universe and the properties of intergalactic space. The light from distant quasars could be used to probe the distribution of matter along the line of sight, revealing the filamentary structure of the cosmic web and the presence of otherwise invisible gas clouds in intergalactic space. Quasar absorption line studies became one of the primary methods for investigating the chemical evolution of the universe and the processes by which heavy elements were distributed throughout cosmic space.

The Cosmic Microwave Background Revelation

The discovery of the cosmic microwave background (CMB) radiation in 1965 by Arno Penzias and Robert Wilson represented perhaps the most profound single discovery in radio astronomy, providing direct observational evidence for the Big Bang theory of cosmic origin and opening new possibilities for studying the early universe in unprecedented detail.

The story of the CMB discovery embodied the serendipitous nature of many great scientific breakthroughs. Penzias and Wilson were not searching for cosmic background radiation when they began their work with Bell Telephone Laboratories' Holmdel Horn Antenna in New Jersey. Instead, they were attempting to measure radio emission from our galaxy with unprecedented precision, hoping to advance our understanding of galactic structure and stellar processes.

However, their observations were plagued by a persistent background noise that seemed to come from all directions in the sky with equal intensity. This noise corresponded to a temperature of about 3.5

degrees above absolute zero and showed no variation with time of day, season, or direction of observation. Penzias and Wilson spent months trying to eliminate possible sources of this unwanted noise, including the famous episode of removing nesting pigeons from their antenna and carefully cleaning away the "white, gray, and black material" the birds had deposited.

The significance of their persistent background signal became clear when Penzias and Wilson learned that Princeton University physicists Robert Dicke, P. James E. Peebles, and their colleagues were actively searching for exactly this type of radiation as a predicted consequence of the Big Bang theory. According to this theory, the universe had begun in an extremely hot, dense state and had been cooling as it expanded. The background radiation that Penzias and Wilson had detected was interpreted as the remnant heat from this primordial fireball, cooled by billions of years of cosmic expansion to just a few degrees above absolute zero.

The discovery of the CMB provided the most compelling evidence for the Big Bang theory and revolutionized cosmology by giving astronomers a direct observational window into the early universe. The radiation that Penzias and Wilson detected had been traveling through space for nearly 13.8 billion years, having been emitted when the universe was only about 380,000 years old and had cooled sufficiently for the first atoms to form. This "surface of last scattering" represented the earliest epoch in cosmic history that could be directly observed, as the universe had been opaque to electromagnetic radiation at earlier times.

Subsequent detailed studies of the CMB revealed tiny temperature fluctuations—variations of only a few parts in 100,000—that represented the primordial density fluctuations from which all cosmic structure would eventually grow. These fluctuations, first detected by NASA's Cosmic Background Explorer (COBE) satellite in the early 1990s and later mapped in exquisite detail by the Wilkinson Microwave Anisotropy Probe (WMAP) and Planck satellites, provided a detailed picture of the conditions in the early universe and enabled precise measurements of fundamental cosmological parameters.

The CMB observations revealed that the universe is geometrically flat, consists of about 5% ordinary matter, 27% dark matter, and 68% dark energy, and began expanding about 13.8 billion years ago. These measurements transformed cosmology from a largely theoretical discipline into a precision science, providing a detailed understanding of cosmic history and evolution that would have been unimaginable before the discovery of the cosmic microwave background.

The Search for Extraterrestrial Intelligence

Radio astronomy's unique capability to detect artificial signals across interstellar distances naturally led to its adoption as the primary tool for searching for evidence of extraterrestrial intelligence. The Search for Extraterrestrial Intelligence (SETI) program, which began in earnest in 1960 with Frank Drake's Project Ozma, represented one of the most ambitious and philosophically significant applications of radio astronomy technology.

The rationale for using radio waves to search for extraterrestrial civilizations was compelling. Radio waves could travel across interstellar distances with minimal absorption or distortion, making them ideal for interstellar communication. Moreover, any technological civilization would likely discover radio astronomy as they developed their understanding of the electromagnetic spectrum, potentially leading them to use radio frequencies for deliberate interstellar signaling or producing detectable radio emissions as byproducts of their technology.

Drake's Project Ozma, conducted using the 85-foot radio telescope at Green Bank Observatory, targeted two nearby sun-like stars—Tau Ceti and Epsilon Eridani—and searched for artificial radio signals at the 21-centimeter hydrogen line frequency. Although the project detected no extraterrestrial signals during its brief observing campaign, it established the methodology and technological foundation for all subsequent SETI research.

The development of SETI programs required significant advances in radio astronomy technology, particularly in the areas of signal processing and data analysis. SETI researchers needed to distinguish potential artificial signals from the vast background of natural radio emission and terrestrial interference. This required sophisticated algorithms capable of identifying narrow-band signals, regular pulses, or other characteristics that might indicate artificial origin.

One of the most famous episodes in SETI history occurred in 1977 when Ohio State University's Big Ear radio telescope detected a strong, narrow-band signal that lasted for 72 seconds and appeared to come from the direction of the constellation Sagittarius. The signal was so unusual that astronomer Jerry Ehman wrote "Wow!" on the computer printout, giving the detection its famous nickname. Despite numerous attempts to detect the signal again, it was never repeated, leaving its nature forever mysterious.

The SETI Institute, founded in 1984, systematized the search for extraterrestrial intelligence and developed increasingly sophisticated search strategies. The Institute's scientists recognized that a comprehensive search would require monitoring millions of stars across a wide range of frequencies, necessitating the development of specialized signal processing systems and search algorithms. Project Phoenix, conducted in the 1990s, represented the most sensitive and comprehensive SETI search ever undertaken, monitoring nearly 800 nearby sun-like stars for artificial signals.

The development of SETI@home in 1999 represented a revolutionary approach to the signal processing challenges facing SETI research. This distributed computing project allowed millions of personal computers around the world to analyze data from the Arecibo radio telescope, creating a supercomputer with processing power that far exceeded any single institution's capabilities. SETI@home demonstrated the potential of citizen science and distributed computing while advancing the technical capabilities of SETI research.

Despite decades of searching, SETI programs have not yet detected confirmed signals from extraterrestrial civilizations. However, the search has contributed significantly to radio astronomy technology development and has helped refine our understanding of the conditions necessary for the development and persistence of technological civilizations. The absence of detected signals has also contributed to discussions about the "Fermi Paradox"—the apparent contradiction between the high probability of extraterrestrial civilizations and the lack of evidence for their existence.

Fast Radio Bursts: The Latest Mystery

The most recent major radio astronomy discovery has been the detection of Fast Radio Bursts (FRBs)—intense pulses of radio emission lasting only milliseconds but releasing as much energy in that brief time as our Sun produces in several days. First discovered in 2007 through analysis of archival data from the Parkes radio telescope in Australia, FRBs have become one of the most intriguing and actively studied phenomena in modern astronomy.

The first FRB, designated FRB 121102, was initially thought to be a unique event—a cosmic flash that occurred once and never repeated. The burst was incredibly intense, producing more radio energy in a few milliseconds than our Sun produces across all wavelengths in an entire day. The signal showed clear signs of having traveled through intergalactic space, exhibiting dispersion effects caused by the small amount of free electrons in the vast emptiness between galaxies.

The discovery of additional FRBs revealed that these events were not unique occurrences but rather representatives of a new class of astronomical phenomena. By 2025, astronomers had detected hundreds of FRBs, revealing a diverse population of sources with different characteristics and behaviors. Some FRBs appeared to be one-time events, while others repeated at irregular intervals. The sources were distributed across the sky, suggesting they originated from cosmological distances rather than within our own galaxy.

The mystery deepened when astronomers began to localize FRBs to their host galaxies using sophisticated interferometric techniques. The first precisely localized FRB was found to originate from a dwarf galaxy billions of light-years away, while subsequent localizations revealed FRBs coming from a diverse range of galaxy types. This diversity suggested that multiple types of astrophysical processes might be capable of producing FRB emissions.

The extreme properties of FRBs—their brief duration, intense energy output, and apparent cosmological origin—posed significant challenges for theoretical explanation. The energy requirements alone ruled out most conventional astrophysical processes, while the brief duration suggested that the emission regions were extremely compact, possibly involving objects no larger than a few hundred kilometers across.

Leading theoretical models for FRB production involved highly magnetized neutron stars called magnetars, which could produce intense bursts of electromagnetic radiation during magnetic field reconfigurations or starquake events. Other models proposed colliding neutron stars, evaporating black holes, or even more exotic possibilities involving cosmic strings or other speculative physics. The diversity of observed FRB properties suggested that multiple mechanisms might be responsible for different types of bursts.

The study of FRBs has opened new possibilities for investigating fundamental physics and cosmology. The dispersion of FRB signals provides a new probe of the intergalactic medium, potentially allowing astronomers to map the distribution of matter between galaxies and study the cosmic web's structure. FRBs may also provide new tests of general relativity and other theories of gravity under extreme conditions.

Mapping the Invisible Universe

Radio astronomy discoveries have revealed that the universe contains vast amounts of material that is completely invisible to optical telescopes but detectable through radio observations. The mapping of neutral hydrogen through the 21-centimeter spectral line has been particularly revolutionary, revealing the distribution of the most common element in the universe and providing insights into galactic structure and evolution that would be impossible to obtain through optical observations alone.

The 21-centimeter line of hydrogen, first detected in 1951, arises from a quantum mechanical transition in neutral hydrogen atoms and provides a unique probe of cool gas in the universe. Unlike optical observations, which are blocked by interstellar dust, 21-centimeter observations can penetrate throughout the galaxy, revealing the distribution and motion of hydrogen gas in unprecedented detail.

Radio observations of hydrogen have revealed that galaxies are embedded in vast rotating disks of gas that extend far beyond the visible stellar components. These observations provided the first evidence for dark matter in galaxies, as the observed rotation curves could not be explained by the gravitational effects of visible matter alone. The flat rotation curves observed in spiral galaxies implied the presence of massive, invisible halos of dark matter extending to many times the optical radius of these systems.

The study of hydrogen in our own galaxy has revealed its spiral structure and has enabled the mapping of star-forming regions throughout the galactic disk. Radio observations have shown that star formation is closely linked to the distribution of molecular hydrogen, which is traced indirectly through observations of carbon monoxide and other molecular species. This connection between gas and star formation has provided crucial insights into the processes that regulate galaxy evolution over cosmic time.

Surveys of hydrogen in nearby galaxies have revealed the diversity of gas content and distribution in different types of systems. Elliptical galaxies generally contain very little neutral hydrogen, while spiral and irregular galaxies show a wide range of gas fractions and distributions. These observations have provided important constraints on models of galaxy formation and evolution, revealing how gas accretion, star formation, and feedback processes shape the properties of galaxies over cosmic time.

The planned Square Kilometre Array (SKA) radio telescope will revolutionize hydrogen astronomy by providing unprecedented sensitivity and resolution for mapping neutral hydrogen throughout the observable universe. The SKA will be capable of detecting hydrogen in galaxies at cosmological distances, enabling studies of galaxy evolution and the cosmic web's structure across most of cosmic history.

Gravitational Wave Astronomy Through Radio Pulses

One of the most sophisticated applications of radio astronomy has been the use of pulsar timing arrays to detect gravitational waves—ripples in spacetime itself predicted by Einstein's general relativity theory. This technique, which monitors the precise arrival times of pulses from multiple millisecond pulsars, has opened an entirely new window on the universe and provided confirmation of one of the most profound predictions of modern physics.

Millisecond pulsars, which rotate hundreds of times per second, serve as extraordinarily precise cosmic clocks whose pulse arrival times can be measured with accuracies of microseconds or better. Gravitational waves passing between pulsars and Earth cause tiny changes in the fabric of spacetime that alter the travel time of pulsar signals by amounts measurable with radio telescopes.

The concept of using pulsars to detect gravitational waves was first proposed in the 1970s, but the required sensitivity was not achieved until the development of highly stable atomic time standards and sophisticated data analysis techniques. Pulsar timing arrays require monitoring dozens of millisecond pulsars for decades to achieve the sensitivity necessary to detect the subtle signatures of gravitational waves.

In 2020, pulsar timing array collaborations around the world announced the detection of a gravitational wave background—a cosmic chorus of gravitational waves from supermassive black hole binaries throughout the universe. This detection represented the first direct observation of gravitational waves using radio astronomy techniques and opened a new frequency window for gravitational wave astronomy that complements the higher-frequency observations made by laser interferometers like LIGO and Virgo.

The gravitational wave background detected through pulsar timing provides unique insights into the population of supermassive black hole binaries in the universe and the processes of galaxy mergers that create these systems. Future pulsar timing array observations may detect individual supermassive black hole mergers and provide new tests of general relativity in the strong-field regime.

Radio Astronomy's Continuing Revolution

As radio astronomy enters the twenty-first century, new discoveries continue to reshape our understanding of the universe. Advanced radio telescopes with unprecedented sensitivity and resolution are revealing previously unknown phenomena while providing new insights into long-studied objects and processes.

The Event Horizon Telescope, a global array of radio telescopes working together as a single Earth-sized instrument, has achieved the seemingly impossible by directly imaging the shadows of supermassive black holes. The first image of the black hole in galaxy M87, released in 2019, provided direct visual confirmation of black hole event horizons and enabled new tests of general relativity in the strongest gravitational fields in the universe.

Advances in radio receiver technology, particularly the development of cryogenic receivers and digital signal processing systems, have dramatically improved the sensitivity and versatility of radio telescopes. These improvements have enabled new types of observations, including real-time detection and localization of transient phenomena like fast radio bursts.

The development of software-defined radio systems has democratized radio astronomy, enabling smaller institutions and even amateur astronomers to conduct meaningful radio observations. These systems allow radio telescopes to be reconfigured dynamically for different types of observations, maximizing their scientific productivity and enabling rapid response to transient phenomena.

Machine learning and artificial intelligence techniques are revolutionizing radio astronomy data analysis, enabling the automated detection of rare phenomena in vast datasets and the extraction of subtle patterns that would be impossible to identify through traditional analysis methods. These techniques are particularly important for next-generation radio surveys that will generate petabytes of data requiring automated processing and analysis.

The Future of Radio Discovery

Looking toward the future, radio astronomy stands poised to make discoveries that may be as revolutionary as pulsars, quasars, and the cosmic microwave background. The Square Kilometre Array, when completed, will provide sensitivity improvements of orders of magnitude over existing radio telescopes, enabling observations that are currently impossible.

The SKA will be capable of detecting radio emission from the first stars and galaxies in the universe, providing direct observations of the cosmic "dark ages" before the first luminous objects formed. These observations will complement infrared observations from space-based telescopes and provide a complete picture of the earliest epochs in cosmic history.

Next-generation radio telescopes will also enable much more sensitive searches for extraterrestrial intelligence, potentially detecting not only deliberately transmitted signals but also the radio emissions from technological civilizations' normal activities. These observations may finally answer the question of whether life and intelligence are common in the universe.

The study of fast radio bursts will continue to evolve as more sophisticated localization techniques enable detailed studies of their host environments and progenitor systems. The eventual identification of FRB progenitors may reveal new classes of astronomical objects or physical processes that are currently unknown.

Radio astronomy will continue to play crucial roles in gravitational wave astronomy, both through pulsar timing arrays and through radio follow-up observations of gravitational wave sources detected by laser interferometers. These multi-messenger observations provide much more complete understanding of cosmic phenomena than any single observational technique can achieve.

The Radio Universe Revealed

The discoveries of radio astronomy have transformed our understanding of the universe from a relatively quiet, stable cosmic arena into a dynamic, violent, and exotic realm filled with phenomena that challenge our understanding of fundamental physics. From Jocelyn Bell's discovery of pulsars to the recent detection of fast radio bursts, radio observations have consistently revealed that the universe is stranger and more wonderful than anyone had imagined.

Radio astronomy has provided unique insights into some of the most extreme phenomena in the universe: stellar remnants so dense that their matter is compressed to nuclear densities, black holes so massive that they contain billions of times the mass of our Sun, and explosions so powerful that they can be detected across billions of light-years. These discoveries have not only advanced our scientific understanding but have also captured the public imagination and inspired new generations of scientists and engineers.

The technological developments driven by radio astronomy research have found applications far beyond astronomy itself. Advances in receiver technology, signal processing, and data analysis have contributed to developments in telecommunications, radar systems, and computing. The techniques developed for radio interferometry have been applied to medical imaging and other fields requiring high-resolution reconstruction from limited data.

Perhaps most importantly, radio astronomy has demonstrated the value of exploring the universe across the full electromagnetic spectrum. Each new wavelength region opened to observation has revealed previously unknown phenomena and provided new insights into cosmic processes. As we continue to develop new observational capabilities and explore new regions of the electromagnetic spectrum, we can be confident that the universe still holds many surprises waiting to be discovered.

The invisible waves that carry radio signals from across the cosmos continue to bring us news of the universe's most dramatic events and exotic objects. Through radio astronomy, we have learned to listen to the universe's radio whispers and thunderous shouts, revealing a cosmic symphony of phenomena that speaks to the fundamental processes that shape the evolution of matter, energy, space, and time throughout the cosmos. The radio universe that Jansky accidentally discovered continues to surprise, inspire, and challenge us, reminding us that the cosmos is far more magnificent and mysterious than our limited human perspective can fully comprehend.

Part V: The Modern Spectrum Battlefield (1980-Present)

Chapter 13: New Technologies, Old Frequencies

The dawn of the 1980s brought with it a technological revolution that would fundamentally alter the radio landscape. For nearly a century, the electromagnetic spectrum had been carefully divided among various users—AM and FM broadcasters, amateur radio operators, maritime services, aviation, and military communications. Each service had its allocated frequencies, and the system, while sometimes contentious, generally worked. But as the final decades of the twentieth century unfolded, a new reality emerged: the appetite for wireless communication would far exceed the available spectrum.

This was the beginning of what engineers and regulators would come to call the "spectrum crunch"—a situation where demand for radio frequencies dramatically outstripped supply. Unlike other natural resources, radio spectrum cannot be manufactured or extracted from the ground. It simply exists, a finite resource governed by the laws of physics, waiting to be used efficiently or squandered carelessly.

The Mobile Revolution Begins

In 1983, Motorola released the DynaTAC 8000X, the first commercially available handheld cellular telephone. Weighing nearly two pounds and costing $3,995 (equivalent to over $12,000 today), it was hardly a mass-market device. But it represented something unprecedented: a wireless communication system that would eventually require enormous swaths of radio spectrum to serve millions, then billions, of users simultaneously.

The cellular concept was elegant in its simplicity. Rather than using a single powerful transmitter to cover a large area—the approach that had worked for broadcasting—cellular systems divided coverage areas into small "cells," each served by a low-power base station. This allowed the same frequencies to be reused in non-adjacent cells, dramatically increasing the system's capacity. But it also meant that cellular operators needed multiple frequency bands to prevent interference between adjacent cells.

The Federal Communications Commission had allocated the 800 MHz band for cellular service, carving out spectrum that had previously been used for other purposes. This marked the beginning of a pattern that would define the spectrum battles of the coming decades: new technologies would not receive virgin frequency allocations, but would instead displace existing users or squeeze into increasingly narrow gaps between established services.

Early cellular systems used analog technology, with each phone call requiring a dedicated radio channel for its entire duration. The inefficiency was staggering—a single phone conversation occupied the same amount of spectrum that could carry hundreds of digital voice calls using modern compression techniques. But in 1983, digital signal processing was still in its infancy, and analog was the only practical option for mobile voice communication.

The Digital Audio Broadcasting Promise

While cellular telephony was taking its first steps, radio broadcasters were grappling with their own technological transition. Edwin Armstrong's FM system had provided superior audio quality compared to AM, but even FM had limitations. Static, multipath interference, and the occasional capture effect—where a weaker signal would suddenly break through a stronger one—reminded listeners that they were still dealing with analog technology.

The solution, many believed, lay in digital audio broadcasting (DAB). The concept was straightforward: convert audio signals into digital data streams, apply error correction coding, and transmit the result using sophisticated modulation schemes. The promised benefits were compelling—CD-quality audio, elimination of static and interference, and the ability to carry additional data services alongside the audio programming.

European engineers took the lead in DAB development, creating the Eureka-147 system in the mid-1980s. The technology was undeniably superior to analog FM, but it faced the same fundamental problem that would plague many digital broadcasting initiatives: it required new spectrum allocations. The Eureka-147 system operated in VHF Band III (174-240 MHz) and the L-band (1452-1492 MHz), frequencies that were either already occupied by other services or considered unsuitable for traditional radio broadcasting.

In the United States, the situation was even more complex. The National Association of Broadcasters (NAB), remembering Edwin Armstrong's disruption of the established AM industry, was skeptical of any technology that might threaten existing FM stations. They insisted that any digital radio system should operate "in-band, on-channel"—meaning that existing FM stations should be able to add digital signals to their current analog transmissions without requiring new frequency allocations.

This constraint would prove to be both a blessing and a curse. It meant that digital radio could be deployed without the massive spectrum reallocation battles that had characterized earlier transitions. But it also imposed severe technical limitations, forcing engineers to squeeze digital signals into the narrow spaces between existing analog FM channels—a challenge akin to building a highway in the median strip of an existing road.

The Rise of Unlicensed Spectrum

One of the most revolutionary developments of the 1980s was the FCC's creation of the Industrial, Scientific, and Medical (ISM) bands for unlicensed operation. The concept was radical: certain frequency bands would be open for anyone to use, provided they followed basic technical rules designed to prevent interference. The ISM bands—at 900 MHz, 2.4 GHz, and 5.8 GHz—had originally been set aside for industrial heating applications, microwave ovens, and medical diathermy equipment. The FCC reasoned that these "polluted" frequencies, already contaminated by intentional radiation from industrial equipment, could safely accommodate low-power communication devices.

The decision unleashed a wave of innovation. Cordless phones quickly migrated to the 900 MHz band, offering better range and sound quality than their earlier 27 MHz and 49 MHz predecessors. Baby monitors, garage door openers, and wireless burglar alarms followed. But the most significant development was still years away: the emergence of wireless local area networks and the technology that would become known as WiFi.

The beauty of unlicensed spectrum was that it democratized innovation. Unlike licensed services, where operators needed FCC approval and often paid substantial fees for spectrum rights, anyone could develop and deploy devices in the ISM bands. This created a fertile environment for experimentation and entrepreneurship, but it also led to a new type of spectrum conflict: the "tragedy of the commons," where multiple users competed for the same frequencies with no central coordination.

The Digital Divide in Broadcasting

As the 1990s dawned, the broadcasting industry found itself caught between two competing visions of the digital future. On one side were the proponents of satellite-delivered digital audio services, who argued that the best way to deliver CD-quality radio was to bypass terrestrial transmitters entirely. On the other were the traditional broadcasters, who insisted that local programming and emergency services required terrestrial transmission infrastructure that satellites could never replace.

The satellite radio advocates had compelling arguments. Satellites could cover the entire continental United States with just a few high-power transmitters, eliminating the need for thousands of local stations and their associated spectrum allocations. The audio quality would be consistently excellent, unaffected by terrain, weather, or multipath interference. And the service could offer hundreds of specialized channels, far exceeding what was possible with traditional broadcasting.

But satellite radio faced its own spectrum challenges. The frequencies best suited for satellite-to-mobile transmission—around 2.3 GHz—were already allocated to other services. Getting new spectrum allocations required not just domestic regulatory approval, but international coordination through the International Telecommunication Union. The process took years and involved complex negotiations with other countries and existing spectrum users.

Meanwhile, terrestrial broadcasters were pursuing their own digital initiatives. The in-band, on-channel approach championed by the NAB had evolved into a technology called HD Radio, which used sophisticated digital signal processing to squeeze digital audio signals into the spaces between existing analog FM channels. The system was technically impressive, but it faced the challenge of backward compatibility—existing analog FM radios couldn't receive the digital signals, meaning that listeners needed new equipment to benefit from the improved audio quality.

The Internet Changes Everything

The emergence of the World Wide Web in the mid-1990s introduced a completely new dimension to radio broadcasting and spectrum usage. For the first time, audio content could be distributed without using radio waves at all. Internet radio stations began appearing, offering everything from professionally produced programming to amateur webcasts operating from bedroom studios.

The implications were profound. Internet radio faced none of the spectrum limitations that constrained traditional broadcasting. A single server could theoretically support unlimited simultaneous listeners, limited only by network bandwidth rather than physics. Geographic coverage was global from day one, without the need for expensive transmitter networks or international frequency coordination.

But internet radio also highlighted the growing importance of wireless internet access, which created new pressure on the radio spectrum. WiFi networks operating in the 2.4 GHz ISM band were multiplying rapidly, and their success created demand for additional unlicensed spectrum. The 5 GHz band, originally allocated for radar and satellite services, became the target of intense lobbying efforts by the computer industry.

The resulting spectrum battles were unlike anything the radio world had seen before. Traditional spectrum users—government radar systems, radio astronomy observatories, and satellite operators—found themselves defending their allocations against an entirely new class of opponent: the information technology industry. These new players had different perspectives on spectrum efficiency, interference tolerance, and the economic value of frequency allocations.

The Cellular Explosion

The introduction of digital cellular technology in the 1990s fundamentally changed the spectrum landscape. Second-generation (2G) systems like GSM and CDMA could handle far more simultaneous conversations than their analog predecessors, but they also enabled entirely new services: text messaging, low-speed data transmission, and the first primitive mobile internet access.

The appetite for spectrum grew exponentially. Cellular operators who had been content with a few dozen megahertz of spectrum in the 1980s now needed hundreds of megahertz to serve growing subscriber bases and increasingly data-hungry applications. The FCC responded by auctioning additional spectrum, but the available frequencies were increasingly marginal—higher in frequency than ideal for mobile coverage, or previously occupied by other services that had to be relocated.

The Personal Communications Services (PCS) auction of 1995 marked a watershed moment in spectrum policy. For the first time, spectrum licenses were sold to the highest bidders rather than awarded based on regulatory merit. The auction raised $7.7 billion for the federal treasury, but it also established a precedent: radio spectrum had become a commodity to be bought and sold like any other natural resource.

This commercialization of spectrum had profound implications for traditional radio broadcasters. AM and FM stations had received their licenses for free, subject to public interest obligations and periodic renewal procedures. Now they watched as cellular operators paid billions for spectrum rights with no content obligations whatsoever. The disparity highlighted fundamental questions about spectrum policy: Should frequency allocations serve the public interest, or should they go to the highest bidder? Could market forces alone ensure efficient spectrum usage?

The Digital Television Transition

The most massive spectrum reallocation in history began in 1997 with the launch of digital television broadcasting. The Advanced Television

Systems Committee (ATSC) standard promised to deliver high-definition video and CD-quality audio while using spectrum more efficiently than analog television. But the transition required unprecedented coordination between broadcasters, equipment manufacturers, and regulators.

Each full-power television station received a second channel for digital broadcasts while continuing to operate their analog transmitters. This doubled the amount of spectrum occupied by television broadcasting in the short term, but it promised to free up enormous amounts of spectrum once the analog transmitters were shut down. The UHF television bands—particularly channels 52-69—were seen as prime real estate for new wireless services.

The transition was scheduled to be completed by 2006, but technical challenges and consumer resistance forced multiple delays. Millions of viewers still relied on analog television reception, and the prospect of making their sets obsolete overnight was politically unpalatable. The final analog shutdown didn't occur until June 2009, twelve years after digital broadcasting began.

The freed spectrum—known as the "digital dividend"—became one of the most contested resources in telecommunications. Cellular operators wanted it for 4G networks, broadcasters sought additional channels for multicasting and mobile television, and public safety agencies needed spectrum for interoperable emergency communications. The resulting battles involved not just technical arguments about propagation characteristics and interference thresholds, but fundamental disagreements about the role of broadcasting in American society.

Software-Defined Radio Revolution

While regulators and industry executives fought over spectrum allocations, a quiet revolution was occurring in radio technology itself. Software-defined radio (SDR) promised to make frequency agility and protocol flexibility routine rather than extraordinary. Instead of building separate hardware for each radio service, SDR systems used digital signal processing and software to implement radio protocols that could be modified, updated, or completely replaced without changing hardware.

The implications for spectrum efficiency were enormous. Traditional radio equipment was designed for a single frequency band and modulation scheme, meaning that spectrum allocations had to account for the technical limitations of existing hardware. SDR systems could adapt to available spectrum in real-time, automatically switching frequencies to avoid interference or taking advantage of temporarily unused channels.

Military and amateur radio operators were among the first to embrace SDR technology, but commercial applications followed rapidly. Cellular base stations began incorporating SDR techniques to support multiple air interface standards simultaneously. WiFi access points started using dynamic frequency selection to avoid radar interference automatically. Even FM radio receivers began using digital signal processing to improve performance in challenging reception environments.

The technology enabled new spectrum sharing concepts that had been impossible with conventional radio equipment. Cognitive radio systems could sense their spectral environment and automatically adjust their operation to minimize interference with other users. Dynamic spectrum access allowed secondary users to opportunistically access licensed spectrum when primary users were inactive.

The Unlicensed Success Story

By the late 1990s, the unlicensed ISM bands had become some of the most intensively used portions of the radio spectrum. WiFi networks were proliferating in offices, homes, and public spaces. Bluetooth enabled short-range communication between computers and peripherals. Cordless phones offered landline-quality audio with wireless convenience. And thousands of other devices—from wireless microphones to remote-control toys—competed for space in these regulatory-free zones.

The success of unlicensed spectrum challenged fundamental assumptions about spectrum management. Traditional regulatory theory held that interference could only be prevented through careful frequency coordination and power limitations enforced by licensing authorities. But the ISM bands demonstrated that well-designed technical standards could enable thousands of devices to coexist without central coordination.

The key was a combination of spread-spectrum technology, adaptive protocols, and statistical multiplexing. Instead of assigning each device a dedicated frequency, unlicensed systems used techniques that allowed multiple transmissions to share the same spectrum simultaneously. Devices learned to tolerate interference, automatically retry failed transmissions, and cooperate to maximize overall system capacity.

This approach had limitations—unlicensed devices generally operated at low power levels and couldn't provide guaranteed quality of service. But for many applications, these constraints were acceptable trade-offs for the flexibility and innovation enabled by unlicensed operation.

The Broadband Revolution

The launch of 3G cellular networks in the early 2000s marked the beginning of the mobile broadband era. For the first time, cellular phones could provide internet access at speeds approaching those of dial-up modems. The implications for spectrum usage were profound—data traffic patterns were fundamentally different from voice communications, with highly variable bandwidth demands and tolerance for delay rather than strict real-time requirements.

Cellular operators found themselves in an arms race for spectrum resources. Voice traffic had predictable busy hours and seasonal variations, but data usage seemed to grow exponentially with no clear upper limit. Email access led to web browsing, which enabled mobile commerce, which created demand for streaming media and social networking applications.

Each generation of cellular technology promised greater spectrum efficiency, but the gains were quickly absorbed by growing traffic demands. The industry began looking beyond traditional cellular bands toward frequencies that had never been considered suitable for mobile services. Millimeter-wave frequencies above 20 GHz offered vast amounts of spectrum, but their propagation characteristics required entirely new network architectures with many more base stations operating at much lower power levels.

Meanwhile, WiFi continued its expansion into new frequency bands. The 5 GHz band provided additional capacity for high-speed wireless LANs, but it also created new interference challenges with weather radar systems and satellite services. The technical solutions required sophisticated database systems to coordinate spectrum sharing between different types of users—a level of complexity that would have been impossible without the computing power and network connectivity that had become routine by the 2000s.

The Convergence Challenge

As the 2000s progressed, the boundaries between different types of wireless services began to blur. Cellular phones evolved into smartphones capable of WiFi connectivity. Tablet computers used both cellular and WiFi networks seamlessly. Even traditional radio broadcasters began offering internet streaming alongside their over-the-air signals.

This convergence created new challenges for spectrum policy. Should a smartphone using WiFi be regulated like a computer network or a radio transmitter? When a tablet computer switches from cellular to WiFi coverage, does it become a different type of device? How should interference be managed between systems that serve similar functions but operate under different regulatory frameworks?

The technical solutions often outpaced regulatory frameworks. Modern wireless devices routinely operate in multiple frequency bands simultaneously, automatically selecting the best available connection without user intervention. Software updates can add support for new air interface standards or modify interference mitigation algorithms. The static frequency allocations that had served the radio industry for decades were increasingly inadequate for managing dynamic, software-controlled wireless systems.

Looking Forward: The Spectrum Crisis

By 2010, it was clear that traditional approaches to spectrum management were reaching their limits. The most valuable frequencies—those below 6 GHz that offered the best combination of coverage and capacity for mobile services—were nearly fully allocated. New wireless applications continued to emerge, but the low-hanging fruit of spectrum policy had been picked.

The solution, many believed, lay in fundamentally rethinking how spectrum was allocated and used. Instead of exclusive licenses for specific services, the future might require dynamic sharing arrangements where multiple users could access the same frequencies under different conditions. Instead of static power limitations, adaptive systems could adjust their transmission characteristics based on real-time interference measurements.

These concepts required technologies that were still in development—cognitive radio systems that could reliably detect other spectrum users, database-driven coordination systems that could manage complex sharing arrangements, and interference mitigation techniques that could enable aggressive frequency reuse. But the alternative was a future where spectrum scarcity would constrain innovation and economic growth.

The stage was set for the next phase of the spectrum wars: a battle not just over which technologies would get which frequencies, but over the fundamental principles that would govern spectrum access in an increasingly wireless world. The easy answers had been exhausted. The hard problems were just beginning.

As we moved into the second decade of the 21st century, one thing was certain: the marriage of new technologies with old frequencies would continue to create challenges that would test the limits of both engineering ingenuity and regulatory wisdom. The invisible waves that had carried humanity's first radio signals more than a century earlier were now the foundation of a global communication infrastructure that touched every aspect of modern life. Managing that precious resource would be one of the defining challenges of the digital age.

Chapter 14: The Internet Age Transformation

In the summer of 1993, as radio engineers around the world grappled with the growing complexity of spectrum allocation and interference management, a computer scientist named Carl Malamud was quietly launching a revolution that would fundamentally challenge every assumption about how audio content should be distributed to audiences. Working from his office, Malamud created something called "Internet Talk Radio"—the world's first internet radio station. Unlike the massive transmitter installations that had defined broadcasting for seven decades, Malamud's station required no towers, no frequency coordination, and no government license. It existed entirely in cyberspace, accessible to anyone with an internet connection and the right software.

The implications of this development would not be immediately apparent to most radio professionals. In 1993, the internet was still largely the domain of universities, government agencies, and technology enthusiasts. The World Wide Web was barely two years old, and most Americans had never heard of email, let alone streaming audio. But Malamud's weekly interviews with computer experts represented something unprecedented in the history of radio: a broadcasting service that operated completely outside the electromagnetic spectrum that had constrained and defined the medium since Marconi's first transmissions.

This was the beginning of the internet age transformation of radio—a period that would see the emergence of entirely new forms of audio distribution, the disruption of traditional broadcasting business models, and the creation of global audiences that transcended the physical limitations of radio waves. More importantly for our story, it marked the beginning of a fundamental shift in how society would think about spectrum scarcity, regulatory authority, and the very definition of "radio" itself.

The Technical Foundation

The technology that made internet radio possible had been developing throughout the 1980s and early 1990s, largely independent of traditional broadcasting. Digital audio compression algorithms, originally developed for storing music on compact discs, were being adapted for transmission over computer networks. The key insight was that audio signals could be converted into streams of digital data, compressed to reduce bandwidth requirements, and then transmitted using the same packet-switched networks that carried email and file transfers.

The technical challenges were formidable. Early internet connections were slow and unreliable compared to modern standards. A typical dial-up modem operated at 9,600 bits per second, barely sufficient for telephone-quality audio and prone to interruptions that could break the audio stream entirely. The internet infrastructure itself was still evolving, with no guarantees about delivery times or data integrity—characteristics that were antithetical to the real-time requirements of broadcasting.

Malamud's solution was elegantly simple: rather than attempting true real-time streaming, Internet Talk Radio offered programs as downloadable files that listeners could retrieve and play at their convenience. This approach bypassed the reliability problems of early internet connections while introducing a concept that would become central to internet media: on-demand access. Listeners were no longer constrained by broadcast schedules or geographic coverage areas. They could access programming whenever they wanted, wherever they had internet access.

The audio quality was modest by today's standards—roughly equivalent to AM radio—but the content was compelling enough to attract a dedicated following among the technology community. Each week, Malamud interviewed prominent figures in computer science and telecommunications, creating an audio archive of the digital revolution as it unfolded. The irony was not lost on observers: the future of radio was being documented in a medium that might ultimately replace radio entirely.

The First Live Streams

While Malamud was pioneering on-demand internet audio, other experimenters were working on true real-time streaming. The first internet concert was broadcast on June 24, 1993, by a band called Severe Tire Damage, performing from a small venue in California. The audio quality was primitive and the audience tiny, but the symbolic significance was enormous: live performance could now be transmitted globally without using any radio spectrum whatsoever.

The technical achievement was remarkable considering the limitations of 1990s internet infrastructure. The concert used experimental software running on high-end workstations connected to the emerging internet backbone. The audio was compressed using algorithms that reduced the data rate to levels that could be handled by the limited bandwidth available, though the result was barely recognizable as music to listeners accustomed to FM radio quality.

More significant was a Rolling Stones concert broadcast in November 1994, which became the "first major cyberspace multicast concert". Mick Jagger opened the show with words that would prove prophetic: "I want to say a special welcome to everyone that's, uh, climbed into the Internet tonight and, uh, has got into the M-bone. And I hope it doesn't all collapse." The M-bone (multicast backbone) was an experimental internet service that allowed a single stream to be efficiently distributed to multiple listeners simultaneously—a crude but effective precursor to modern content delivery networks.

The Rolling Stones concert demonstrated both the potential and the limitations of internet broadcasting. The global reach was unprecedented—listeners from around the world could experience a live concert without traveling or relying on local radio stations to carry the broadcast. But the technical quality was marginal, and the complexity of accessing the stream limited the audience to technology enthusiasts with sophisticated equipment and high-speed internet connections.

Traditional Radio Enters Cyberspace

The first traditional radio station to embrace internet broadcasting was WXYC, a college station in Chapel Hill, North Carolina, which began streaming its FM signal online on November 7, 1994. The technical setup was characteristically improvised: an FM radio receiver connected to a computer running experimental software at the University of North Carolina's SunSite facility. The stream was announced to a small community of internet users through email and early web sites, creating the first hybrid between traditional radio broadcasting and internet distribution.

WXYC's innovation was significant because it demonstrated that internet streaming could complement rather than replace traditional radio. The station continued its normal FM broadcasts while simultaneously offering its programming to a global internet audience. This dual-mode operation would become the standard approach for most radio stations, allowing them to expand their reach without abandoning their existing listeners or their valuable spectrum licenses.

Other college radio stations quickly followed WXYC's lead. WREK in Atlanta launched its own streaming service on the same day, using custom software called CyberRadio1. These early adopters were primarily non-commercial educational stations that had the technical expertise and experimental freedom to explore new distribution methods. Commercial broadcasters, constrained by complex licensing agreements and conservative business models, were slower to embrace internet streaming.

The quality of these early streams was limited by both technology and economics. Audio compression algorithms of the mid-1990s required significant processing power and produced artifacts that were noticeable even to casual listeners. Internet bandwidth was expensive and unreliable, making it difficult to maintain consistent audio quality. Most importantly, the audience for internet radio was still tiny—limited to the small percentage of the population with both internet access and the technical knowledge to configure streaming audio software.

The Birth of Internet-Only Radio

While traditional broadcasters experimented with simulcasting their existing programming online, entrepreneurs began creating radio stations that existed only on the internet. Scott Bourne founded NetRadio.com in 1995 as the world's first internet-only radio network. Unlike the experimental efforts that had preceded it, NetRadio.com was designed as a commercial venture that would eventually compete directly with traditional broadcasters.

NetRadio.com represented a fundamental shift in radio business models. Without the need for expensive transmitter sites or spectrum licenses, internet-only stations could be launched with minimal capital investment. They could offer specialized programming that would never be commercially viable on traditional radio, serving niche audiences scattered across wide geographic areas. Most importantly, they could operate without the regulatory constraints that governed traditional broadcasting—no content restrictions, no public interest obligations, and no limits on commercial time.

The success of NetRadio.com inspired a wave of internet radio entrepreneurship. By the late 1990s, thousands of internet-only stations were broadcasting everything from traditional music formats to highly specialized programming that served communities defined by interest rather than geography. Live365.com became the first platform to offer hosted internet radio services, allowing anyone to create and operate an online station without technical expertise or significant investment.

This democratization of radio had profound implications for the traditional broadcasting industry. For the first time in radio's history, the barriers to entry for audio programming had been essentially eliminated. Anyone with a computer and an internet connection could potentially reach a global audience, competing directly with established radio stations that had invested millions of dollars in spectrum licenses and transmission facilities.

The Streaming Revolution

The introduction of RealAudio in 1995 marked the beginning of mainstream internet radio. Developed by RealNetworks, RealAudio was the first streaming audio technology designed for consumer use rather than experimental applications. It offered reasonable audio quality over dial-up connections and could be accessed through standard web browsers with minimal technical complexity.

RealAudio's impact was immediate and transformative. Within months of its release, hundreds of radio stations began offering RealAudio streams of their programming. The technology was still primitive—installation was cumbersome, audio quality was inconsistent, and streaming frequently failed due to network congestion—but it demonstrated that internet radio could serve general audiences rather than just technology enthusiasts.

The success of RealAudio attracted major technology companies to the streaming market. Microsoft developed Windows Media Player with integrated streaming capabilities, while Nullsoft created Winamp, which became the preferred audio player for a generation of internet users. These free software tools dramatically reduced the technical barriers to internet radio, making streaming audio accessible to mainstream computer users.

The proliferation of streaming software created new opportunities for content creators and distributors. Music labels began experimenting with internet-only releases and promotional streams. Independent artists discovered they could distribute their music globally without the gatekeeping function of traditional radio programmers. Most importantly, listeners gained unprecedented control over their audio experience—they could choose from thousands of stations worldwide, create custom playlists, and access programming on demand rather than according to broadcast schedules.

Challenging the Spectrum Paradigm

The rapid growth of internet radio in the late 1990s forced regulators and industry observers to reconsider fundamental assumptions about spectrum management and media regulation. For decades, radio policy had been based on the premise that electromagnetic spectrum was a scarce resource that required careful allocation and intensive oversight to prevent interference and ensure public service. Internet radio operated in an entirely different paradigm where distribution capacity was theoretically unlimited and barriers to entry were minimal.

The contrast was particularly stark when comparing the costs and complexity of traditional radio licensing with internet radio deployment. A typical FM radio station required an initial investment of hundreds of thousands of dollars for transmitter equipment, tower construction, and studio facilities, plus ongoing fees for spectrum licensing and regulatory compliance. An internet radio station could be launched with equipment costing less than $1,000 and operated from any location with internet access.

This disparity raised uncomfortable questions for traditional broadcasters and regulators. If internet radio could provide the same services as traditional radio without using spectrum, what justified the complex regulatory framework that governed traditional broadcasting? Should internet radio stations be subject to the same content regulations and public interest obligations as spectrum-licensed broadcasters? How should advertising and commercial revenue be regulated in a medium where geographic boundaries were meaningless?

The regulatory response was typically cautious and fragmented. The Federal Communications Commission maintained that internet radio fell outside its jurisdiction because it didn't use radio spectrum, while other agencies struggled to determine how existing media regulations should apply to online audio services. This regulatory vacuum allowed internet radio to develop with minimal government oversight, creating a laboratory for innovation that would have been impossible under traditional broadcasting regulations.

The Napster Revolution

The launch of Napster in 1999 fundamentally altered the relationship between internet technology and the music industry, with profound implications for radio broadcasting. While Napster was primarily a file-sharing service rather than a streaming platform, it demonstrated the disruptive potential of internet-based music distribution and foreshadowed challenges that would ultimately transform radio's role in music discovery and promotion.

Napster's peer-to-peer architecture allowed users to share music files directly with each other, effectively creating a vast library of recorded music that could be accessed on demand without payment to artists or record labels. The service grew explosively, reaching millions of users within months and creating the first large-scale alternative to traditional music distribution channels.

For radio broadcasters, Napster represented both a threat and an opportunity. On one hand, it competed directly with radio's traditional role as a music discovery medium—listeners could now access virtually any song instantly rather than waiting to hear it on the radio. On the other hand, it demonstrated the enormous appetite for on-demand music access, suggesting that radio might need to evolve beyond traditional programming models to remain relevant.

The music industry's response to Napster was swift and decisive. Legal action eventually shut down the original Napster service, but the concept of internet-based music distribution had been permanently established. Legitimate services like iTunes and later Spotify would offer legal alternatives that combined Napster's convenience with proper compensation for artists and record labels.

The Napster controversy also highlighted the complex intellectual property issues surrounding internet radio. Traditional broadcasters paid licensing fees to performing rights organizations for the music they played, but internet radio operated in a regulatory gray area where existing licensing agreements might not apply. The resolution of these issues would require years of negotiation and litigation, ultimately creating new licensing frameworks specifically for internet radio operations.

The Rise of Streaming Services

The early 2000s saw the emergence of sophisticated streaming music services that combined elements of internet radio with on-demand access and personalization features. Pandora, launched in 2000, pioneered the concept of algorithmic radio programming, using the Music Genome Project to analyze songs and create personalized radio stations based on listeners' preferences. This approach represented a fundamental departure from traditional radio programming, which relied on human program directors to select music for broad demographic groups.

Pandora's innovation lay in its ability to create virtually unlimited programming variations tailored to individual listeners. Rather than offering a handful of stations designed to appeal to large audiences, Pandora could generate thousands of unique "stations" based on specific artists, songs, or musical characteristics. This mass customization was impossible with traditional radio technology but became routine in the internet environment where individual streams could be created and managed automatically.

The success of Pandora inspired competitors and imitators. Last.fm developed sophisticated social networking features that allowed users to share musical preferences and discover new artists through their social connections. Spotify, launched in Europe in 2006, combined streaming radio with on-demand access to millions of songs, creating a hybrid service that challenged both traditional radio and music sales.

These services represented the maturation of internet radio from experimental technology to mainstream media platform. By the mid-2000s, streaming music services were attracting millions of regular users and generating significant advertising revenue. More importantly, they were beginning to influence music consumption patterns and artist promotion strategies in ways that traditional radio had dominated for decades.

The Podcasting Revolution

The development of podcasting in the early 2000s created another parallel to traditional radio that operated entirely outside the electromagnetic spectrum. Podcasts were essentially on-demand audio programs that could be automatically downloaded and synchronized to portable devices, allowing listeners to access spoken-word content with the convenience of recorded music but the topical relevance of live radio.

The term "podcast" was coined by journalist Ben Hammersley in 2004, combining "iPod" and "broadcast" to describe this new form of audio distribution. The technology built on existing internet infrastructure and RSS (Really Simple Syndication) protocols that had been developed for web content management. Listeners could subscribe to podcast "feeds" that would automatically deliver new episodes as they became available.

Podcasting democratized spoken-word programming in ways that paralleled internet radio's impact on music broadcasting. Anyone could create and distribute a podcast without technical expertise or significant investment. The format was particularly well-suited to niche programming that would never be commercially viable on traditional radio—specialized topics, unusual perspectives, and extended discussions that couldn't fit into conventional broadcast formats.

The growth of podcasting was explosive. By 2005, major media companies were launching podcast versions of their radio programming, while thousands of independent creators were developing original content specifically for the podcast format. Apple's integration of podcast support into iTunes provided a mainstream distribution platform that reached millions of potential listeners.

For traditional radio, podcasting represented both competition and opportunity. Talk radio hosts discovered they could extend their reach beyond their local broadcast markets by offering podcast versions of their shows. News organizations could provide in-depth coverage that complemented their on-air programming. But podcasting also enabled new competitors who could develop loyal audiences without the overhead and regulatory constraints of traditional broadcasting.

Satellite Radio's Spectrum Strategy

While internet radio was disrupting traditional broadcasting through software and networks, satellite radio was attempting to achieve similar goals using advanced spectrum-based technology. XM Satellite Radio and Sirius Satellite Radio both launched commercial services in the early 2000s, offering CD-quality audio programming delivered directly to mobile receivers from high-power satellites.

Satellite radio represented a massive investment in spectrum and infrastructure. Both companies had paid hundreds of millions of dollars for spectrum licenses in the 2.3 GHz band and deployed sophisticated satellite constellations specifically designed for mobile audio broadcasting. The services offered hundreds of channels, including commercial-free music, sports, news, and talk programming that was impossible to provide through traditional terrestrial broadcasting.

The satellite radio business model was fundamentally different from both traditional broadcasting and internet radio. Unlike traditional radio, which was supported by advertising to local audiences, satellite radio relied on subscription fees from individual listeners. Unlike internet radio, which could be accessed with general-purpose computer equipment, satellite radio required specialized receivers and service subscriptions.

The competition between satellite radio and internet radio highlighted different approaches to spectrum utilization and business models. Satellite radio offered superior audio quality, nationwide coverage, and programming that was specifically designed for mobile listening. Internet radio provided greater variety, lower costs, and the flexibility to access programming on multiple devices.

The eventual merger of XM and Sirius in 2008 reflected the challenges facing spectrum-based approaches to audio distribution in an increasingly internet-centric world. While satellite radio maintained a loyal subscriber base, it struggled to compete with the convenience and economics of internet-based services that continued to improve in quality and reduce in cost.

The Convergence Era

By 2010, the boundaries between internet radio, traditional radio, satellite radio, and on-demand music services had become increasingly blurred. Smartphones with both cellular and WiFi connectivity could access any of these services seamlessly, while automotive entertainment systems began integrating internet streaming alongside traditional AM/FM reception. The concept of "radio" was evolving to encompass any audio programming delivered in real-time, regardless of the underlying technology.

This convergence created new challenges for regulators, content creators, and technology developers. Should a smartphone app that streams traditional radio programming be regulated as internet content or broadcasting? When a car's entertainment system automatically switches between terrestrial radio, satellite radio, and internet streaming based on signal quality, which regulatory framework applies? How should advertising and content restrictions be applied to services that might be accessed through multiple delivery mechanisms simultaneously?

The technical solutions often outpaced regulatory frameworks. Modern audio services routinely used adaptive streaming that automatically adjusted quality based on available bandwidth, hybrid systems that combined multiple delivery methods for reliability, and recommendation algorithms that personalized programming based on listening history. These capabilities were impossible with traditional broadcasting technology but became expected features of internet-based services.

The economic implications were equally profound. Internet radio services could measure audience engagement with unprecedented precision, providing detailed analytics about listener behavior, geographic distribution, and demographic characteristics. This data enabled more sophisticated advertising strategies and content optimization, while also raising privacy concerns about the collection and use of personal information.

The Global Transformation

The internet age transformation of radio extended far beyond technological change to encompass social, cultural, and economic shifts that would have been impossible in the spectrum-limited world of traditional broadcasting. Internet radio enabled global communities organized around shared interests rather than geographic proximity. A jazz enthusiast in rural Iowa could access specialized programming from stations in New York, London, or Tokyo, while an expatriate could maintain connection to their homeland through internet streams of local radio stations.

This globalization of audio content challenged traditional concepts of local programming and community service that had been central to broadcast radio regulation. If listeners could access unlimited programming from anywhere in the world, what obligation did local stations have to serve their immediate communities? How should advertising be regulated when audiences might be scattered across multiple countries and time zones? What role should local content requirements play in an environment where "local" was increasingly meaningless?

The democratizing effects of internet radio were particularly pronounced in countries with restrictive media policies or limited spectrum allocations. Independent journalists, political dissidents, and cultural minorities could create internet radio services that reached global audiences without government licensing or approval. These services often provided alternative perspectives and programming that would never be allowed on traditional broadcast media.

The transformation also enabled new forms of audience participation and interaction. Internet radio listeners could communicate with hosts and other listeners through chat rooms, social media, and email, creating communities around shared musical tastes or topical interests. Some services allowed listeners to vote on programming choices or contribute content directly, blurring the lines between producers and consumers of audio content.

Looking Forward: The Spectrum Liberation

As we entered the second decade of the 21st century, the internet age transformation of radio was fundamentally changing how society thought about spectrum allocation and media regulation. Internet-based audio services had demonstrated that high-quality, diverse programming could be delivered to global audiences without using radio spectrum at all. This raised fundamental questions about the continued necessity of traditional broadcasting and the regulatory frameworks that had governed it for nearly a century.

The implications extended beyond radio to encompass television, telephony, and other spectrum-intensive services that were increasingly being delivered through internet infrastructure. The massive investments in spectrum licenses and transmission facilities that had characterized the broadcasting industry for decades might become unnecessary if internet delivery could provide superior service at lower cost.

But the transition was not without challenges. Internet radio still depended on internet infrastructure that might not be available in emergency situations where traditional broadcasting provided essential public safety services. Rural and remote areas with limited internet connectivity might be poorly served by internet-only audio services. And the concentration of internet services among a few major technology companies raised concerns about media diversity and democratic discourse that spectrum-based broadcasting regulations had been designed to address.

The stage was set for continued evolution in the relationship between internet technology and spectrum-based services. The lessons learned from internet radio's disruption of traditional broadcasting would inform policy decisions about spectrum allocation, media ownership, and content regulation for decades to come. The invisible waves that had carried radio signals for more than a century would continue to play important roles, but they would increasingly share the audio landscape with the packet-switched networks that had made the internet age transformation possible.

The revolution that began with Carl Malamud's modest experiment in 1993 had fundamentally altered the media landscape, creating new possibilities for content creation, audience engagement, and global communication that transcended the physical limitations of electromagnetic spectrum. The future of audio media would be shaped not just by advances in radio technology, but by the continued evolution of internet infrastructure and the creative ways that entrepreneurs and artists would use these tools to connect with audiences around the world.

Part VI: The Future Spectrum (Present-2050)

Chapter 15: 5G, IoT, and the Crowded Airwaves

In the spring of 2019, as wireless engineers gathered at the World Radiocommunication Conference in Sharm El Sheikh, Egypt, they faced a challenge that would have been unimaginable to the radio pioneers of a century earlier. The electromagnetic spectrum that had once seemed infinite—vast enough to accommodate a handful of experimental stations and maritime safety services—was now being carved into ever-smaller slices to serve billions of connected devices. The conference delegates were tasked with allocating approximately 17 GHz of millimeter-wave spectrum for the fifth generation of mobile networks, representing one of the largest spectrum assignments in the history of radio.

This massive allocation was not merely an incremental upgrade from previous mobile generations. 5G represented a fundamental shift in how wireless networks would be conceived, deployed, and used. Unlike its predecessors, which had been designed primarily for human communication, 5G was engineered from the ground up to serve a world where machines would outnumber human users by orders of magnitude. The Internet of Things (IoT) promised to connect everything from automotive sensors to industrial machinery, from smart city infrastructure to agricultural monitoring systems. Each of these applications would require its own slice of the increasingly crowded electromagnetic spectrum.

The irony was profound: just as internet radio and streaming services were demonstrating that high-quality audio could be delivered without using radio spectrum at all, the wireless industry was demanding unprecedented amounts of spectrum to serve applications that previous generations of radio engineers had never imagined. The collision between these trends would define the spectrum battles of the 2020s and beyond, forcing regulators, engineers, and society as a whole to reconsider fundamental assumptions about how the invisible resource of radio frequencies should be allocated and used.

The 5G Spectrum Architecture

The technical requirements that drove 5G spectrum allocations were unlike anything the wireless industry had previously encountered. Where 4G networks had been designed to serve millions of smartphones with relatively predictable traffic patterns, 5G needed to accommodate billions of IoT devices with vastly different requirements for latency, bandwidth, and reliability. A self-driving car required ultra-low latency communication to avoid accidents, while a soil moisture sensor might transmit tiny amounts of data just once per day. Industrial robotics demanded guaranteed reliability, while consumer streaming applications could tolerate occasional interruptions.

The solution was a three-tier spectrum strategy that would span nearly the entire usable radio frequency range. Low-band 5G, operating in frequencies below 1 GHz, would provide wide-area coverage similar to traditional cellular networks. Mid-band 5G, using frequencies from 1.7 to 4.7 GHz, would offer a balance between coverage and capacity for urban and suburban deployments. High-band 5G, operating in millimeter-wave frequencies from 24 to 71 GHz, would deliver gigabit speeds over short distances in dense urban environments.

This multi-tier approach represented a fundamental departure from previous mobile network architectures, which had relied primarily on a single frequency band for each generation of technology. The complexity was staggering: a single 5G network might simultaneously operate dozens of different frequency bands, each with its own propagation characteristics, interference patterns, and regulatory constraints. The technical challenge of coordinating these diverse spectrum resources would require advances in antenna technology, signal processing, and network management that pushed the boundaries of what was technically feasible.

The frequency allocations themselves told the story of spectrum scarcity. Low-band 5G was forced to share spectrum with existing 4G networks through a process called "dynamic spectrum sharing," where the same frequencies could be used by both technologies simultaneously. Mid-band allocations required the displacement of incumbent services, including satellite communications, radio astronomy, and military radar systems. High-band millimeter-wave spectrum, while relatively unoccupied, presented propagation challenges that would require entirely new network architectures with thousands of small cells operating at much higher densities than previous generations.

The Millimeter-Wave Frontier

The decision to extend mobile communications into millimeter-wave frequencies above 24 GHz represented one of the most ambitious spectrum ventures in radio history. These frequencies had been largely unexplored for mobile applications because their propagation characteristics seemed incompatible with traditional cellular network designs. Millimeter waves were easily blocked by buildings, absorbed by rain, and attenuated by atmospheric gases. Their range was measured in hundreds of meters rather than kilometers, and they required precise antenna alignment for reliable communication.

But millimeter waves offered something that lower frequencies could not: vast amounts of contiguous spectrum. The frequency range from 24 to 71 GHz contained more available bandwidth than all lower-frequency mobile allocations combined. In a world where spectrum efficiency had become the paramount concern for wireless network design, millimeter waves represented the last frontier for capacity expansion.

The technical solutions required to make millimeter-wave mobile communication practical pushed multiple technologies to their limits. Massive MIMO (Multiple Input Multiple Output) antenna systems with 64 to 256 elements became standard equipment for millimeter-wave base stations, using sophisticated beamforming techniques to focus radio energy precisely where it was needed. Advanced materials science enabled the development of phased array antennas that could electronically steer beams without mechanical movement. Machine learning algorithms optimized beam patterns in real-time based on user locations and interference conditions.

The deployment model for millimeter-wave 5G was fundamentally different from traditional cellular networks. Instead of a few large cell towers covering wide areas, millimeter-wave systems required thousands of small cells mounted on street poles, building facades, and indoor locations. Each small cell covered only a few hundred meters, but collectively they created a dense mesh network capable of delivering multi-gigabit data rates to mobile devices. The infrastructure requirements were enormous—a typical urban deployment might require ten times more base stations than 4G networks to achieve comparable coverage.

The economic implications were equally challenging. While millimeter-wave spectrum was often available at lower cost than prime mid-band frequencies, the infrastructure costs were substantially higher. Mobile operators found themselves investing billions of dollars in dense fiber networks to connect thousands of small cells, while struggling to develop business models that could justify the enormous capital expenditures. The result was a selective deployment strategy that focused on high-value locations like sports stadiums, airports, and dense urban cores where the capacity benefits could command premium pricing.

The Internet of Things Explosion

The emergence of IoT applications as a major driver of spectrum demand represented a paradigm shift from the human-centric communication models that had dominated wireless development for decades. Where previous mobile generations had focused on serving human users with relatively predictable communication patterns, IoT promised to connect tens of billions of devices with radically different requirements for spectrum usage, power consumption, and network reliability.

The diversity of IoT applications defied traditional spectrum allocation models. Smart city sensors monitoring air quality required occasional transmission of small data packets with minimal power consumption. Industrial automation systems needed guaranteed low-latency communication with extremely high reliability. Agricultural monitoring networks required wide-area coverage with years of battery life. Connected vehicles demanded real-time coordination with infrastructure and other vehicles. Each application category required different spectrum characteristics, but all competed for the same finite frequency resources.

The technical solution was the development of specialized IoT communication protocols designed for spectrum efficiency rather than peak data rates. Narrowband IoT (NB-IoT) compressed device communications into narrow 180 kHz channels that could be deployed within existing cellular spectrum allocations. Long Range (LoRa) networks used spread-spectrum techniques to enable low-power devices to communicate over distances of several kilometers using minimal spectrum resources. Sigfox developed ultra-narrowband technology that could support thousands of devices within a single traditional radio channel.

But the proliferation of IoT technologies also created new spectrum management challenges. Unlike licensed cellular networks with centralized coordination, many IoT networks operated in unlicensed spectrum bands where interference management relied on protocol design rather than regulatory control. The 2.4 GHz ISM band, already crowded with WiFi networks, Bluetooth devices, and microwave ovens, became even more congested as billions of IoT sensors joined the frequency sharing environment.

The scale of IoT deployment was unprecedented in the history of wireless communication. Industry forecasts predicted 50 to 100 billion connected IoT devices by 2030, compared to roughly 10 billion smartphones and tablets. Each device might transmit only occasionally, but the aggregate spectrum demand was enormous. More challenging was the diversity of deployment environments—IoT devices would operate in factories, forests, urban centers, and remote agricultural areas, each with different propagation characteristics and interference conditions.

Urban Spectrum Warfare

The concentration of wireless activity in urban environments created spectrum congestion that pushed traditional frequency coordination methods to their breaking point. A single city block in Manhattan might contain hundreds of WiFi networks, dozens of cellular base stations, thousands of Bluetooth devices, and millions of IoT sensors, all competing for access to the same electromagnetic spectrum. The interference environment was so complex that traditional propagation models became nearly useless for predicting network performance.

The technical response was the development of cognitive radio technologies that could automatically detect and adapt to local interference conditions. Advanced algorithms analyzed the spectrum environment in real-time, identifying unused frequencies and adjusting transmission parameters to minimize interference with other users. Machine learning systems optimized network performance based on historical patterns of spectrum usage, automatically shifting resources to match changing demand.

But cognitive radio also introduced new regulatory challenges. Traditional spectrum management relied on static frequency assignments that could be monitored and enforced by regulatory agencies. Cognitive systems that automatically changed frequencies based on local conditions made it nearly impossible to verify compliance with technical rules or resolve interference complaints. The FCC and other regulatory bodies struggled to develop new frameworks that could accommodate dynamic spectrum access while maintaining the predictability that had enabled decades of interference-free operation.

The urban spectrum crunch also drove innovations in network architecture that fundamentally challenged traditional approaches to wireless system design. Heterogeneous networks combined multiple radio technologies—cellular, WiFi, and specialized IoT protocols—in integrated systems that could automatically route traffic based on application requirements and spectrum availability. Software-defined networking enabled network operators to dynamically reconfigure spectrum allocations based on real-time demand patterns.

The economic pressures of urban spectrum scarcity led to the emergence of spectrum sharing business models that would have been unthinkable in earlier eras of wireless development. Citizens Broadband Radio Service (CBRS) in the 3.5 GHz band enabled multiple operators to share the same frequencies through database-coordinated dynamic protection of incumbent users. Private networks for enterprises began operating in shared spectrum, reducing the demand for exclusive licensed frequencies while enabling customized wireless solutions for specific applications.

The Global Spectrum Patchwork

The internationalization of wireless technology created unprecedented challenges for spectrum coordination across national boundaries. While earlier generations of mobile networks had used broadly similar frequency bands worldwide, 5G deployments revealed a complex patchwork of national spectrum policies that reflected different priorities, incumbent services, and regulatory philosophies.

European regulators emphasized harmonized spectrum allocations that would enable economies of scale in equipment manufacturing and roaming between countries. The 3.4 to 3.8 GHz band became the primary European allocation for 5G mid-band services, cleared through coordinated relocation of satellite services and military radar systems. China focused on similar frequencies but with different boundary conditions, creating equipment compatibility challenges for global device manufacturers. The United States developed a unique approach that combined mid-band sharing arrangements with aggressive millimeter-wave allocations, reflecting the political difficulty of relocating incumbent government users from prime spectrum.

These international differences had profound implications for equipment manufacturers and network operators. A single smartphone needed to support dozens of different frequency bands to achieve global compatibility, adding complexity and cost to device design. Network equipment manufacturers faced the challenge of developing products that could be cost-effectively adapted to different regional spectrum plans while maintaining economies of scale in production.

The fragmentation of global spectrum allocations also created opportunities for geopolitical competition in wireless technology standards. Chinese manufacturers like Huawei and ZTE developed 5G equipment optimized for spectrum bands used in China and developing countries, while European and American manufacturers focused on different frequency ranges. The result was a bifurcation of global wireless infrastructure that reflected broader tensions about technology sovereignty and security.

International coordination mechanisms developed over decades of radio regulation struggled to cope with the pace and complexity of 5G spectrum requirements. The International Telecommunication Union's traditional processes for global spectrum harmonization required years of preparation and negotiation, while wireless technology evolved on timescales measured in months. Regional bodies like the European Conference of Postal and Telecommunications Administrations (CEPT) and the Inter-American Telecommunication Commission (CITEL) became more important for rapid spectrum coordination, but their regional focus sometimes conflicted with the global scale of wireless technology markets.

Interference in the Modern Age

The dramatically increased density of wireless systems in the 5G era created interference challenges that dwarfed anything previously encountered in radio engineering. Traditional interference analysis had assumed relatively simple scenarios with a few high-power transmitters operating in predictable patterns. The 5G environment featured millions of low-power devices with adaptive transmission characteristics, creating interference patterns that changed continuously and unpredictably.

The technical solutions required fundamental advances in signal processing and antenna technology. Interference cancellation algorithms could identify and suppress unwanted signals in real-time, allowing receivers to operate successfully even in severely congested spectrum environments. Advanced antenna systems with hundreds of elements could form precise beam patterns that minimized interference while maximizing desired signal strength. Adaptive modulation schemes automatically adjusted transmission parameters based on local interference conditions.

But the proliferation of adaptive and cognitive radio technologies also created new categories of interference that were difficult to predict or resolve. When multiple systems attempted to use cognitive techniques simultaneously, their adaptive algorithms could interact in unexpected ways, creating interference patterns that emerged from the interaction of multiple well-intentioned systems. The complexity of these interactions often exceeded the ability of traditional engineering analysis to predict or remedy.

The regulatory response to modern interference challenges required new approaches to technical standards and enforcement. Traditional type acceptance procedures that tested equipment under idealized laboratory conditions were inadequate for evaluating systems that would adapt their behavior based on real-world deployment conditions. New testing methodologies needed to account for the statistical nature of interference in dense deployment environments while ensuring that equipment would not cause unacceptable interference under worst-case conditions.

The international nature of modern wireless systems complicated interference resolution when problems occurred. A interference issue in one country might originate from equipment manufactured in another country, using protocols developed by international standards bodies, and deployed by operators with global business models. Traditional mechanisms for resolving interference disputes between national administrations were poorly suited to addressing problems that transcended national boundaries and involved multiple stakeholders with different regulatory relationships.

The Millimeter-Wave Promise and Reality

By 2025, the reality of millimeter-wave 5G deployment had both validated and challenged the promises made by wireless industry advocates. The technology successfully delivered multi-gigabit data rates in carefully controlled environments, enabling applications like augmented reality gaming and ultra-high-definition video streaming that had been impossible with previous wireless technologies. Major sporting venues and transportation hubs deployed millimeter-wave networks that provided exceptional user experiences for dense crowds of connected devices.

But the propagation limitations of millimeter-wave frequencies proved more challenging than many early advocates had anticipated. The promise of ubiquitous gigabit connectivity required infrastructure investments that exceeded initial projections by substantial margins. Indoor coverage required dedicated small cell installations that were often cost-prohibitive for residential and small business applications. Weather sensitivity, particularly rain attenuation, created reliability challenges that required sophisticated network redundancy and fallback mechanisms.

The selective deployment of millimeter-wave 5G created a new form of digital divide based on spectrum access rather than network availability. Premium locations with dense millimeter-wave coverage offered dramatically superior wireless performance, while areas served only by low-band 5G provided connectivity that was often only marginally better than 4G networks. This disparity had implications for economic development, education, and social equity that extended far beyond technical considerations.

The equipment ecosystem for millimeter-wave 5G evolved more slowly than initially expected. Consumer devices required sophisticated antenna systems and power-hungry RF circuits that impacted battery life and manufacturing costs. Many applications that had been touted as driving forces for millimeter-wave adoption—autonomous vehicles, industrial automation, and smart city systems—proved to have requirements that could be met by lower-frequency 5G technologies with less complex infrastructure.

IoT and the Spectrum Commons

The explosive growth of IoT devices created new models for spectrum sharing that challenged traditional concepts of exclusive licensing and interference protection. Unlike mobile phone users, who expected consistent performance and were willing to pay premium prices for spectrum access, IoT applications often involved low-value devices with minimal quality of service requirements. The economic model of exclusive spectrum licensing became problematic when applied to sensors that might transmit just a few bytes of data per day.

The solution was the expansion of unlicensed and shared spectrum models that allowed multiple operators and applications to coexist in the same frequency bands. The Citizens Broadband Radio Service in the 3.5 GHz band pioneered dynamic protection systems that enabled secondary users to access spectrum as long as they didn't interfere with incumbent services. Similar sharing frameworks were developed for other frequency bands, creating a spectrum commons that could accommodate the diverse and unpredictable requirements of IoT deployments.

But spectrum sharing also created new technical and business challenges. IoT device manufacturers needed to develop products that could operate reliably in unpredictable interference environments without the guaranteed quality of service that licensed spectrum traditionally provided. Network operators had to design systems that could gracefully degrade performance when spectrum became congested rather than failing catastrophically when interference exceeded acceptable levels.

The regulatory framework for IoT spectrum sharing required fundamental changes to traditional approaches to equipment authorization and interference protection. Instead of protecting individual licensed users from any detectable interference, sharing systems needed to manage aggregate interference levels across multiple users and services. This required sophisticated measurement and enforcement techniques that could account for the statistical nature of interference in shared spectrum environments.

The international coordination of IoT spectrum sharing was particularly challenging because different countries adopted incompatible approaches to shared spectrum management. Some nations emphasized strict technical rules that limited device capabilities but provided predictable interference protection. Others adopted more flexible approaches that enabled innovative applications but created uncertainty about interference levels and quality of service.

The Future Spectrum Crisis

As the 2020s progressed, it became increasingly clear that traditional approaches to spectrum allocation were inadequate for the demands of an increasingly connected world. The fundamental problem was mathematical: the total amount of radio spectrum below 100 GHz was finite, but the demand for wireless connectivity appeared to be growing exponentially. Even with dramatic improvements in spectrum efficiency, the projected demand for wireless services by 2030 would exceed the supply of usable spectrum by substantial margins.

The technical solutions under development pushed the boundaries of physics and engineering. Massive MIMO antenna systems with thousands of elements could focus radio energy with unprecedented precision, enabling aggressive frequency reuse in dense deployments. Advanced signal processing techniques could extract useful information from signals that were buried below the noise floor using traditional detection methods. Orbital angular momentum and other exotic modulation schemes promised to increase spectral efficiency beyond what had been thought theoretically possible.

But these technical advances came with substantial complexity and cost penalties that limited their practical deployment. Massive MIMO systems required sophisticated calibration and control systems that increased infrastructure costs and power consumption. Advanced signal processing techniques worked well in laboratory conditions but often failed when deployed in real-world environments with unpredictable interference and propagation conditions.

The economic implications of the spectrum crisis were becoming apparent in wireless infrastructure investments and consumer pricing. Operators were paying increasingly high prices for spectrum licenses while simultaneously investing billions in network densification to achieve acceptable performance with limited spectrum resources. These costs were inevitably passed on to consumers through higher service prices and device costs, creating affordability challenges that threatened to limit the benefits of advanced wireless technology to high-income users and applications.

The policy response to spectrum scarcity required fundamental reconsideration of regulatory frameworks that had been developed when spectrum was relatively abundant. Traditional models of exclusive licensing and interference protection were increasingly incompatible with the diversity of applications and the intensity of competition for spectrum access. New approaches needed to balance the efficiency benefits of market mechanisms with the equity and reliability concerns that had historically justified regulatory oversight.

Looking Toward 2030

The trajectory of wireless technology development suggested that the spectrum challenges of the mid-2020s were just the beginning of a transformation that would reshape how society used and managed radio frequencies. The convergence of artificial intelligence, edge computing, and advanced materials science promised to enable wireless applications that would dwarf current spectrum demands while potentially offering solutions to current technical limitations.

Artificial intelligence was beginning to revolutionize spectrum management through predictive algorithms that could optimize network performance based on anticipated rather than historical usage patterns. Machine learning systems could identify subtle patterns in interference and propagation that enabled more aggressive frequency reuse while maintaining acceptable quality of service. AI-driven network optimization promised to extract dramatically more capacity from existing spectrum allocations, but it also required computing resources and energy consumption that created new infrastructure challenges.

The development of 6G wireless systems was already beginning to influence spectrum policy discussions, even as 5G deployments were still in their early stages. Initial research suggested that 6G would require access to frequencies above 100 GHz, extending wireless communications into portions of the electromagnetic spectrum that had never been used for mobile applications. The technical challenges were formidable—atmospheric absorption increased dramatically at these frequencies, and conventional semiconductor technology struggled to generate and detect signals efficiently.

But perhaps the most significant development was the growing recognition that spectrum scarcity was not just a technical problem but a fundamental constraint on economic and social development. The wireless infrastructure that had become essential for modern commerce, education, and social interaction was approaching physical limits that could not be overcome through incremental improvements in technology or regulation.

The solutions being developed for this challenge would require unprecedented cooperation between industry, government, and international organizations. The spectrum wars that had characterized radio development for more than a century were evolving into a more complex challenge of managing a global commons that was essential for human prosperity but inherently limited by the laws of physics.

The invisible waves that had carried humanity's first radio signals were now the foundation of a global nervous system that connected billions of people and devices in real-time communication networks of unprecedented complexity and capability. Managing this precious resource would require the best efforts of engineers, policymakers, and society as a whole to ensure that the benefits of wireless technology could be sustained and equitably distributed in an increasingly connected world.

The future of radio spectrum would be determined not just by technical innovation, but by the wisdom and cooperation of the global community in managing one of the most valuable and finite resources of the modern age. The decisions made in the coming decade would shape the trajectory of wireless technology and its impact on human society for generations to come.

Chapter 16: The Final Frontier - Space Communications

On October 4, 1957, as the Soviet Union launched Sputnik 1 into orbit, radio operators around the world heard something extraordinary: a simple, regular beeping sound transmitted from space on frequencies of 20.005 and 40.01 MHz. These signals, barely more sophisticated than Marconi's early experiments, marked the beginning of a new era in radio communication. For the first time in human history, radio waves were not just traveling across the Earth's surface but were being transmitted from beyond our atmosphere, opening what would become known as the "final frontier" of spectrum utilization.

The beeping of Sputnik was more than a technological achievement or a geopolitical triumph in the Cold War. It represented the first practical demonstration that radio spectrum was not just a terrestrial resource but could serve as humanity's connection to the vastness of space. In the decades that followed, space communications would evolve from simple satellite beacons to complex global networks that would transform how we understand our world, explore the universe, and connect with each other.

The Early Space Race Frequencies

The Soviet Union's decision to use relatively low frequencies for Sputnik was not merely a technical convenience but a strategic one. The 20 MHz and 40 MHz bands were well within the range of standard shortwave radio equipment available to amateur radio operators worldwide. This meant that virtually anyone with a basic receiver could detect the satellite's signal, creating a global audience for the Soviet achievement and demonstrating the power of radio as a tool of international communication and influence.

The Americans responded quickly. When Explorer 1 became the first U.S. satellite to reach orbit on January 31, 1958, it carried beacons on 108.00 and 108.03 MHz—frequencies that had been specified by an international committee for the International Geophysical Year (IGY) as the standard for scientific satellites. This frequency lay just above the terrestrial FM broadcast band (88 to 108 MHz) and just inside the civil aviation band (108 to 136 MHz), representing a carefully considered technical decision that reflected growing international cooperation in space research.

The contrast between the Soviet and American approaches highlighted fundamental differences in early space communication strategies. The Soviets had chosen to ignore international frequency coordination efforts, using lower frequencies that were more accessible to amateur observers but less suitable for scientific data transmission. The Americans, by contrast, had worked within the emerging international framework, selecting frequencies that would minimize interference with existing services while maximizing the scientific utility of the satellite's transmissions.

These early frequency choices set the stage for decades of technical and political negotiation over how the radio spectrum would be shared between terrestrial and space-based users. The simple beeping of Sputnik had demonstrated that space was not beyond the reach of radio waves, but it had also revealed that the electromagnetic spectrum would become a contested resource in humanity's expansion beyond Earth.

The Birth of Space Communication Bands

As satellite technology advanced beyond simple beacons to sophisticated communication and scientific missions, the need for dedicated frequency bands became increasingly apparent. The early 1960s saw the first organized efforts to coordinate space communication frequencies on an international scale. In October 1963, the Extraordinary Administrative Radio Conference (EARC) in Geneva, Switzerland, made history by assigning specific frequency allocations for space services for the first time.

This conference, organized by the International Telecommunication Union (ITU), represented a critical moment in the evolution of space communications. The EARC was entirely dedicated to space communications matters, following up on discussions that had emerged after the 1959 World Administrative Radio Conference. The allocations established at this conference would form the foundation for space communication systems for decades to come.

The conference established several key frequency bands for space communications:

- VHF Band (136-138 MHz): This band was heavily used by many different types of satellites in the early years of space communication. While limited in bandwidth, VHF offered reliable propagation characteristics and compatibility with existing terrestrial radio equipment.

- S Band (2.025-2.3 GHz): As satellite technology advanced, the need for higher bandwidth drove the adoption of S band frequencies. The higher frequencies offered greater data capacity while still providing reasonable propagation characteristics for space-to Earth communication.

- X Band (7.145-7.190 GHz uplink, 8.400-8.450 GHz downlink): By the 1970s, X band had become the standard for many scientific and military satellite communications, offering even greater bandwidth for data-intensive missions.

- Ka Band (34.2-34.7 GHz uplink, 31.8-32.3 GHz downlink): The highest frequency band allocated in the early years, Ka band would eventually become critical for high-bandwidth satellite communications, though its adoption was initially limited by technical challenges.

- These allocations reflected a careful balance between technical considerations and political negotiation. The lower frequency bands were already crowded with terrestrial services, making them difficult to clear for space communications. The higher frequency bands offered greater bandwidth but presented significant technical challenges due to atmospheric absorption and the limitations of early semiconductor technology.

The Lincoln Experimental Satellites and Beyond

The development of space communication technology advanced rapidly through a series of experimental satellite programs, most notably the Lincoln Experimental Satellite (LES) series developed by the U.S. military. LES-9, launched on March 14, 1976, represented a quantum leap in space communication capabilities and would go on to become the longest continuously operating satellite in U.S. history, and likely the world.

LES-9 was the last and most technologically sophisticated of the Lincoln Experimental Satellites. It was designed to address critical Cold War concerns about the robustness of military communications in the face of potential jamming or other interference. The satellite demonstrated several first-time space communication technologies that would shape future systems:

- Ka-band frequencies: LES-9 operated in the 37 to 38 GHz range, providing many GHz of bandwidth for spread-spectrum modulation that could effectively reject interference. This was particularly significant because these frequencies were experimental at the time, as radio frequency bands in these regions had not yet been assigned by U.S. and international agencies.

- Onboard demodulation and digital switching: These technologies allowed the satellite to reject uplink interference and enable flexible routing of traffic among multiple coverage areas.

- Multiple frequency crossbanding: This technique added flexibility to communications connectivity by allowing signals to be received on one frequency band and transmitted on another.

- Code-division multiple access (CDMA): LES-9 demonstrated CDMA techniques that allowed a large number of asynchronous, bursty-traffic users to more efficiently utilize bandwidth. This technology would later become the foundation for cellular telephone standards.

The success of LES-9 and its companion satellite LES-8 proved the feasibility of expanding satellite communications into the very wideband Ka-band frequencies and incorporating resiliency into communications among mobile users. These technologies would eventually ensure that U.S. commanders could stay connected to their forces worldwide and that civilians could enjoy high-speed communication services.

The Technical Challenge of Space Propagation

Space communications presented unique propagation challenges that were fundamentally different from terrestrial radio. While terrestrial radio waves could be reflected by the Earth's atmosphere or follow the Earth's surface through ground wave propagation, space communications had to account for the vacuum of space, the Earth's atmosphere, and the vast distances involved.

The first major challenge was signal attenuation. Radio waves follow the inverse fourth power of distance in space communications (compared to inverse second power for terrestrial communications), meaning that signal strength decreases dramatically with distance. A satellite in geosynchronous orbit (35,786 km above Earth) experiences signal attenuation approximately 100,000 times greater than a terrestrial communication over 1 km.

The Earth's atmosphere presented additional challenges. As radio waves passed through the atmosphere, they encountered absorption by atmospheric gases, particularly at higher frequencies. The absorption characteristics of the atmosphere created natural "radio windows" where signals could more easily reach space:

- VHF and UHF bands (30-3000 MHz): These frequencies experience minimal atmospheric absorption but require large antennas for space communication due to their long wavelength.

- C band (4-8 GHz): This band experiences some atmospheric absorption but is still suitable for satellite communications, particularly for weather satellites.

- Ku band (12-18 GHz): This band is heavily used for satellite television broadcasting but experiences increased rain attenuation.

- Ka band (26.5-40 GHz): This band experiences significant atmospheric absorption, particularly during rain, but the high frequencies enable small antennas and high data rates.

The highest frequencies useful for space communication are limited by the absorption of microwave energy by the atmosphere. As frequency increases above 30 GHz (the beginning of the millimeter wave band), atmospheric gases absorb increasing amounts of power, so the power in a beam of radio waves decreases exponentially with distance from the transmitting antenna. At 30 GHz, useful communication is limited to about 1 km, but as frequency increases the range at which the waves can be received decreases. In the terahertz band above 300 GHz, the radio waves are attenuated to zero within a few meters due to the absorption of electromagnetic radiation by the atmosphere (mainly due to ozone, water vapor and carbon dioxide), which is so great that it is essentially opaque to electromagnetic emissions, until it becomes transparent again near the near-infrared and optical window frequency ranges.

These propagation characteristics forced space communication engineers to make careful trade-offs between bandwidth, antenna size, and reliability. The selection of frequency bands for specific space missions became a complex optimization problem that considered not just technical performance but also regulatory constraints, international coordination requirements, and the availability of suitable equipment.

The Space Communication Ecosystem

By the 1980s, space communications had evolved from experimental satellite beacons to a complex ecosystem of specialized networks serving different purposes. This ecosystem included:

- The NASA Space Network (SN): A system of geosynchronous communications satellites (known as the Tracking and Date Relay Satellite System or TDRS) that provided near-continuous communication with low Earth orbit spacecraft. The network initially operated in S-band (1960s), then moved to X-band in the 1990s, with Ka-band beginning to be used in the 21st century. Many spacecraft have dual frequency capability (S/X and later X/Ka) to ensure compatibility with evolving ground infrastructure.

- The NASA Deep Space Network (DSN): A global network of large radio antennas that support interstellar spacecraft missions. The DSN uses S-band, X-band, and increasingly Ka-band for communication with spacecraft throughout the Solar System and beyond.

- Military Satellite Communications (MILSATCOM): A series of specialized networks including the UHF Follow On (UHF-10), the

Wideband Global SATCOM (WGS), and the Advanced EHF systems, each operating in different frequency bands to serve specific military requirements.

- International Space Communication Systems: Systems operated by other space agencies including the European Space Tracking network (ESTRACK), the Chinese Deep Space Network, and the Soviet/Russian network that supported missions like the International Space Station.

- Satellite Television and Radio: Commercial services like SiriusXM satellite radio and various satellite television services that deliver entertainment content directly to consumers.

- Satellite Internet and Mobile Services: Systems like Iridium, Global Star, and Inmarsat that provide voice and data services to mobile users worldwide.

Each of these networks operated in carefully coordinated frequency bands to avoid interference with terrestrial services and with each other. The coordination required international cooperation through the ITU and national regulatory bodies to ensure that the finite resource of radio spectrum could support the growing demands of space communication.

The International Space Frequency Challenge

The international coordination of space frequencies presented unique challenges that were more complex than terrestrial spectrum management. Unlike terrestrial services where interference typically occurred between adjacent geographic areas, space communications could potentially interfere with services across entire countries or even multiple nations.

The ITU's Radio Communication Sector (ITU-R) became the primary international organization for coordinating space frequencies. The ITU established a complex framework of regulations and procedures for allocating space frequencies, including:

- The Radio Communication Conference (RCC): A regular international conference where nations could debate and vote on frequency allocations for space services.

- The International Frequency Registration Board (IFRB): A technical organization that maintained the International Frequency Registration Board (MIFR) database of all frequency assignments worldwide.

- The Space Frequency coordination process: A detailed technical process for coordinating satellite networks to avoid interference with existing services.

The coordination process was particularly challenging for geosynchronous satellites, which occupied specific "orberal" locations above the equator. Each geosynchronous satellite required not just a frequency allocation but also a specific location in the geosynchronous arc to avoid interference with other satellites. This created a situation where "space" in the geosynchronous orbit was as valuable as spectrum itself, with nations and companies competing for prime locations above their home countries.

The complexity of international space frequency coordination was highlighted by the "space race" of the 1990s, when dozens of nations and companies sought to deploy satellite constellations for mobile communications and internet access. The Iridium system, Global Star, and other constellations required careful coordination to avoid interference with each other and with existing services. The process took years of technical analysis and negotiation, demonstrating that the invisible resource of radio spectrum had become a critical strategic concern for nations and companies operating in space.

The Commercialization of Space Communications

The late 1980s and 1990s saw the commercialization of space communications, as private companies began to develop satellite systems for commercial applications. This shift created new challenges for spectrum management, as commercial interests competed with government and scientific users for access to valuable frequency bands.

The most dramatic example was the development of satellite constellations for mobile communications and internet access. Iridium, launched in the late 1990s, deployed a network of 66 satellites operating in the L-band (1616-1626.5 MHz) to provide global voice and data services. Global Star followed with a similar system operating in the S-band (2483.5-2500 MHz). These systems required massive investments in satellite manufacturing, launch services, and ground infrastructure, but they also required careful spectrum coordination to avoid interference with existing services.

The commercial satellite industry faced significant challenges in spectrum allocation. Many of the most valuable frequency bands were already allocated to government or scientific services, forcing commercial operators to use less ideal frequencies or develop innovative spectrum sharing techniques. The L-band, for example, was already heavily used by military communications, air traffic control, and scientific research, making it difficult to clear for commercial satellite services.

The situation was further complicated by the emergence of terrestrial wireless services that sought to use the same frequencies as satellite systems. The 2 GHz band, for example, became the subject of intense debate as terrestrial 3G cellular networks sought to use frequencies that had been allocated to satellite services. This "spectrum squeeze" created difficult policy questions about whether satellite services should have primary or secondary rights to specific frequency bands.

The Modern Space Communication Landscape

By the early 21st century, space communications had become an essential part of global infrastructure, serving a wide range of applications from scientific research to commercial services. The modern space communication landscape includes several key elements:

- The Deep Space Network: NASA's global network of large radio antennas that support interstellar spacecraft missions. The DSN has evolved from S-band to X-band and now increasingly to Ka-band, with each transition providing higher data rates for scientific missions. The shift to higher frequencies has enabled missions like the New Horizons spacecraft to transmit high-definition video from the edge of the Solar System.

- The Space Network: NASA's Tracking and Date Relay Satellite System (TDRS) has evolved through multiple generations, with the most advanced satellites (TDRS-12 and beyond) capable of operating in both S-band and Ka-band. This dual-band capability ensures compatibility with legacy spacecraft while enabling higher data rates for new missions.

- Satellite Internet: Systems like Iridium, Global Star, and more modern constellations like Starlink and One Web are creating a new generation of satellite internet services. These systems operate in various frequency bands, with Starlink using Ka-band (18.8-19.3 GHz and 28.1-28.6 GHz) and V-band (37.5-42.0 GHz) for its high-speed services.

- Satellite Radio: Services like SiriusXM continue to operate in the S-band (2320-2345 MHz), providing entertainment and information to millions of consumers worldwide.

- Satellite Television: Modern satellite television services operate primarily in the Ka-band (17.7-21.2 GHz downlink) and increasingly in the V-band for high-definition and ultra-high-definition services.

- Military SATCOM: Modern military satellite communications systems operate across multiple bands, with strategic systems using extremely high-frequency uplinks (around 44 GHz) and Ka-band downlinks (near 20 GHz), while tactical systems use a variety of frequency bands to meet specific operational requirements.

The modern space communication landscape is characterized by increasing frequency congestion and the need for sophisticated spectrum sharing techniques. As more satellites are launched and more services compete for limited spectrum resources, the technical and regulatory challenges of space communications continue to grow.

The Spectrum Squeeze and Interference Challenge

The growing demand for space communication services has created what engineers call the "spectrum squeeze"—a situation where the demand for radio frequencies far outstrips the available supply. This squeeze is particularly pronounced in the most valuable frequency bands for space communications, where the physical characteristics of the frequencies make them ideal for specific applications.

The L-band (1-2 GHz), for example, is highly valuable for satellite communications because it experiences minimal atmospheric absorption while still providing reasonable bandwidth. This band is used by satellite radio services, satellite phone systems, and increasingly by terrestrial 5G networks. The competition for L-band spectrum has led to complex sharing arrangements and technical innovations to minimize interference between different services.

The S-band (2-4 GHz) is another highly contested frequency range. It is used by satellite television, satellite radio, weather satellites, and terrestrial 4G and 5G networks. The proximity of satellite services to terrestrial mobile networks has created challenging interference scenarios, particularly as terrestrial networks increase their power levels to serve growing data demands.

The Ka-band (26.5-40 GHz) has become increasingly important for high-bandwidth satellite communications, but it also experiences significant atmospheric absorption, particularly during rain. This absorption limits the reliability of Ka-band services but also has the benefit of reducing interference between geosynchronous satellites, as the Earth's atmosphere attenuates signals from satellites that are not directly above the receiver.

The most significant interference challenges in modern space communications come from:

- Satellite-to-satellite interference: As the number of satellites in orbit increases, particularly in low Earth orbit constellations, the potential for interference between different satellite systems becomes more significant.
- Satellite-to-terrestrial interference: The growing use of satellite frequencies by terrestrial networks, particularly for 5G services, has created new interference scenarios where terrestrial signals can interfere with satellite reception.
- Space-to-space interference: For missions beyond Earth orbit, interference between different space agencies' spacecraft can become a concern, particularly for sensitive scientific measurements.
- Spectrum sharing challenges: The traditional model of exclusive frequency allocations is increasingly incompatible with the growing demand for spectrum, forcing regulators to develop new sharing models that can accommodate multiple users in the same frequency bands.

The Future of Space Communications

As we approach the mid-21st century, space communications are entering a new era characterized by unprecedented growth in satellite deployments, increasingly sophisticated spectrum sharing techniques, and the emergence of new applications that will further challenge the limits of radio spectrum.

The most significant development is the proliferation of satellite constellations in low Earth orbit. Systems like Starlink, One Web, and future constellations promise to provide global broadband internet access, but they also require massive amounts of spectrum to serve billions of potential users. These constellations operate in multiple frequency bands, including:

- Ku-band (12-18 GHz): Traditional satellite internet frequencies that are now being used by new constellations.
- Ka-band (26.5-40 GHz): High-bandwidth frequencies that enable gigabit-speed internet services from space.
- V-band (40-75 GHz): The next frontier for satellite communications, offering enormous bandwidth but challenging propagation characteristics.
- E-band (71-76 GHz and 81-86 GHz): Even higher frequencies that may be used for future satellite communications, particularly for inter-satellite links.

The technical challenges of operating in these higher frequency bands are significant. As frequency increases, atmospheric absorption becomes more severe, particularly during rain. The higher frequencies also require more precise antenna alignment and are more sensitive to physical blockage. However, the enormous bandwidth available at these frequencies may be essential for serving the growing demand for satellite internet services.

The Interference Challenge of Massive Satellite

The deployment of massive satellite constellations has created new interference challenges that are fundamentally different from previous space communication systems. Traditional geosynchronous satellites operated in carefully coordinated frequency plans with relatively few satellites in orbit. Modern constellations, by contrast, may include thousands of satellites operating in multiple frequency bands, creating complex interference scenarios that are difficult to predict and manage.

The most significant interference concerns include:

- Uplink interference: As more users connect to satellite internet services, the potential for interference between user uplinks becomes more significant. This is particularly challenging for systems that use frequency reuse across multiple satellites.

- Downlink interference: The dense deployment of satellites in low Earth orbit means that multiple satellites may be simultaneously transmitting to the same geographic area, creating potential interference between different satellites' signals.

- Satellite-to-satellite interference: For constellations with thousands of satellites, the potential for interference between different satellites in the same system becomes a significant concern.

- Impact on radio astronomy: The proliferation of satellite constellations has created significant concerns for radio astronomy, as satellite transmissions can interfere with sensitive scientific measurements. The 10.6-10.7 GHz band, for example, is critical for radio astronomy but is also used by some satellite systems.

The technical solutions to these interference challenges include:

- More sophisticated beamforming: Advanced antenna systems that can form highly focused beams to minimize interference with other users.

- Cognitive radio techniques: Systems that can automatically detect and avoid interference in real-time.

- Spectrum sharing protocols: Technical standards that enable multiple systems to share the same frequency bands without interference.

- International coordination: More sophisticated frameworks for coordinating satellite networks on a global scale.

The Space Spectrum Commons

The growing demand for space communication services is driving the development of new spectrum management models that go beyond traditional exclusive licensing. These models include:

- Spectrum sharing: Systems that can dynamically share frequency bands based on real-time usage patterns. This approach is particularly valuable for space communications, where satellite coverage areas may not always be fully occupied.

- Spectrum commons: Frequency bands that are open for use by multiple operators under technical rules designed to prevent interference. The success of unlicensed spectrum in terrestrial communications (e.g. WiFi in the 2.4 GHz and 5 GHz bands) has inspired similar approaches for space communications.

- Spectrum markets: Systems that enable the dynamic allocation of spectrum resources based on market mechanisms, allowing operators to trade spectrum usage rights in real-time.

- Cognitive spectrum access: Systems that can automatically detect unused spectrum and use it without interference to primary users.

These new models are particularly important for space communications because the traditional approach of exclusive frequency allocations is increasingly incompatible with the growing demand for spectrum resources. The limited availability of suitable frequency bands for space communications means that more efficient sharing mechanisms are essential for enabling the continued growth of space-based services.

The Future: Beyond Radio Waves

As we approach the limits of what is possible with radio frequency communications, space agencies and commercial companies are beginning to explore alternative communication technologies that could complement or even replace traditional radio systems.

The most promising alternative is optical communications, which uses infrared or near-infrared light to transmit data between spacecraft and ground stations. NASA's demonstration of optical communications with the Lunar Laser Communication Demonstration (2013) and the Laser Communications Relay Demonstration (2021) has opened a new frontier in space communications.

The benefits of optical communications include:

- Vastly higher bandwidth: Laser communication systems can achieve data rates that are orders of magnitude higher than radio frequency systems.

- Narrower beamwidth: The highly focused nature of optical beams means that less power is required for communication, and the narrow beams are less likely to cause interference with other systems.

- Spectrum availability: The optical spectrum is not regulated in the same way as radio frequencies, potentially eliminating spectrum allocation challenges.

- More efficient use of spacecraft resources: Laser communication systems can be more compact and power efficient than radio frequency systems, particularly for high data rate applications.

The challenges of optical communications include:

- More precise alignment requirements: The narrow beamwidth that is an advantage for interference mitigation also means that spacecraft and ground stations must maintain extremely precise alignment.

- Vulnerability to atmospheric conditions: Weather conditions like rain, fog, and even atmospheric turbulence can severely impact optical communications.

- More complex ground infrastructure: Receiving optical signals from space typically requires specialized ground stations with adaptive optics to counter atmospheric effects.

- Current technological limitations: While optical communication technology is rapidly maturing, it is not yet as reliable or well-proven as traditional radio frequency systems.

The future of space communications is likely to be hybrid, with radio frequency systems continuing to serve applications where reliability and all- weather operation are critical, while optical systems handle high-bandwidth applications where the higher data rates justify the additional complexity.

Looking Toward 2050

As we project the future of space communications toward 2050, several trends are becoming increasingly clear:

- Spectrum scarcity will intensify: The demand for space communication services will continue to grow, particularly as satellite internet services expand and new applications emerge. The most valuable frequency bands will become increasingly congested, requiring more sophisticated spectrum sharing techniques.

- The rise of multi-band systems: Future satellite systems will likely operate across multiple frequency bands simultaneously, using the most suitable band for each specific application and environment.

- The integration of space and terrestrial networks: The boundaries between space-based and terrestrial communication networks will continue to blur, with integrated systems that seamlessly handoff between satellite and terrestrial connections.

- The emergence of space-based spectrum management: As the number of satellites in orbit increases, the need for real-time spectrum coordination will become critical. This may lead to the development of space-based spectrum management systems that can coordinate frequency usage across multiple satellite networks.

- The expansion into higher frequencies: The continued development of semiconductor technology will enable the practical use of higher frequency bands for space communications, particularly in the millimeter wave and terahertz ranges.

- The growth of optical communications: While radio frequency systems will remain essential for many applications, optical communications will become increasingly important for high-bandwidth applications, particularly for inter-satellite links and missions beyond Earth orbit.

- The development of quantum communication: Quantum key distribution and other quantum communication technologies may become important for space-based applications requiring ultra-secure communications.

The future of space communications will be shaped by the same fundamental constraints that have governed radio since Marconi's first experiments: the laws of physics that determine how electromagnetic waves interact with the environment. But it will also be shaped by human ingenuity in developing new technologies and regulatory frameworks that can maximize the utility of this invisible resource.

As we continue to explore and utilize space, the radio spectrum will remain our essential connection to the final frontier. The beeping of Sputnik that first announced humanity's entry into space was transmitted on frequencies that had been used for terrestrial communication for decades. Today, the complex signals that connect our global satellite infrastructure continue to use the same electromagnetic spectrum, but with technologies that would have been unimaginable to those early pioneers.

The challenge for the coming decades will be to ensure that this precious resource can continue to serve humanity's growing need for space communication while also enabling the scientific discovery and technological innovation that will shape our future in space. The invisible waves that carry our signals to and from space are not just a technical resource but a fundamental enabler of humanity's expansion beyond Earth—a connection to the final frontier that will continue to transform our world for generations to come.

Chapter 17: Beyond the Horizon - Future Technologies

As we stand at the threshold of 2025, looking toward the horizon of 2050 and beyond, the electromagnetic spectrum that has carried humanity's radio signals for more than a century is poised for transformations that would seem like magic to Guglielmo Marconi. The simple spark-gap transmitter that sent the first wireless signals across the Atlantic in 1901 has evolved into a global nervous system of unprecedented complexity and capability. Yet we are only at the beginning of a new chapter in the radio story—one that will be written not just with electromagnetic waves, but with quantum phenomena, artificial intelligence, and technologies that exist today only in the realm of theoretical possibility.

The radio waves that carry our voices, our data, and our dreams remain governed by the same fundamental laws of physics that Maxwell first described in his equations. But our understanding of how to manipulate, control, and transcend the limitations of the electromagnetic spectrum has advanced to the point where we can envision a future where spectrum scarcity itself may become obsolete. This final chapter explores the technologies and paradigms that will define radio communication in the decades to come—innovations that will push the boundaries of what we thought possible and perhaps fundamentally redefine what we mean by "radio" itself.

The Intelligence Revolution

The integration of artificial intelligence into radio systems represents perhaps the most profound transformation since the invention of the transistor. By 2025, AI-powered radio systems are already demonstrating capabilities that seemed impossible just a decade earlier. These systems can automatically adjust frequencies to avoid interference, predict network congestion before it occurs, and optimize signal parameters in real-time based on changing environmental conditions. But this is merely the beginning of what artificial intelligence will accomplish in radio communications.

The next generation of AI-powered radio systems will possess what engineers call "cognitive radio intelligence"—the ability to understand and adapt to their electromagnetic environment with a sophistication that approaches human-level reasoning. These systems will not simply react to interference; they will anticipate it, predict optimal frequency assignments hours or days in advance, and coordinate with millions of other radio devices to create self-organizing networks of unprecedented efficiency.

The implications extend far beyond technical performance improvements. AI-powered radios will be capable of automatically negotiating spectrum sharing agreements between competing users, resolving interference disputes without human intervention, and even creating new communication protocols optimized for specific environmental conditions. Emergency response radios equipped with AI will automatically prioritize critical communications during disasters, filter out irrelevant traffic, and provide real-time translation between different languages and communication protocols.

Perhaps most remarkably, these intelligent radio systems will begin to exhibit emergent behaviors—capabilities that arise from the interaction of millions of individual devices rather than being explicitly programmed. Just as the internet developed capabilities that were never envisioned by its original designers, AI-powered radio networks will discover new ways to use the electromagnetic spectrum that human engineers have never considered.

The convergence of AI with Software-Defined Radio (SDR) technology will create radios that can completely reconfigure themselves on demand. A single device will be capable of operating as a cellular phone, a WiFi router, a satellite communicator, or a radar system, switching between these functions instantly based on user needs and spectrum availability. The traditional concept of dedicated radio equipment for specific purposes will give way to universal communication devices that can adapt to any requirement.

The Quantum Leap

While artificial intelligence will revolutionize how we use the electromagnetic spectrum, quantum technologies promise to transcend the fundamental limitations that have constrained radio communication since its inception. Quantum communication systems, already demonstrated in laboratory environments, will begin commercial deployment in the 2030s, offering capabilities that classical radio systems cannot match.

Quantum key distribution (QKD) will provide theoretically unbreakable encryption for radio communications, making it impossible for adversaries to intercept and decode messages even with unlimited computational resources. Unlike classical encryption, which relies on mathematical complexity, quantum encryption is protected by the fundamental laws of physics—any attempt to eavesdrop on a quantum communication automatically introduces detectable changes to the signal.

But quantum technologies will offer more than just enhanced security. Quantum radar systems, which use entangled photons to detect objects, will be capable of operating effectively even in environments with intense electronic jamming. Traditional radar can be defeated by transmitting noise in the same frequency band, but quantum radar creates correlations between transmitted and received signals that cannot be replicated by jammers. This capability will be particularly valuable for military applications and civilian air traffic control in contested electromagnetic environments.

Quantum sensing technologies will enable radio systems to detect and measure electromagnetic fields with precision that approaches theoretical limits. Quantum magnetometers, already under development, can detect magnetic field variations that are billions of times smaller than Earth's magnetic field. Applied to radio communications, these sensors will enable new techniques for direction finding, interference source identification, and propagation measurement that will dramatically improve spectrum efficiency.

The most speculative but potentially transformative quantum technology is quantum computing applied to spectrum management. Quantum computers excel at solving optimization problems that are intractable for classical computers, such as finding optimal frequency assignments for millions of simultaneous users across multiple geographic regions. As quantum computers mature in the 2040s, they may enable real-time optimization of global spectrum usage that makes current interference problems obsolete.

Beyond Electromagnetic Waves

The electromagnetic spectrum, vast as it may seem, is ultimately finite. The laws of physics impose absolute limits on how much information can be transmitted through radio waves in any given frequency band. As we approach these theoretical limits in the most valuable portions of the spectrum, researchers are exploring entirely new communication paradigms that transcend electromagnetic radiation altogether.

Magnetic field communication represents one such paradigm. Unlike electromagnetic waves, magnetic fields can penetrate through materials that block radio waves, enabling communication through soil, rock, and seawater with minimal attenuation. Magnetic communication systems, already used for specialized applications like communication with submarines, will expand to serve underground mining operations, geological surveys, and buried infrastructure monitoring.

The frequency spectrum for magnetic communication extends far below the traditional radio spectrum, into the extremely low frequency (ELF) range where wavelengths are measured in thousands of kilometers. While this limits data transmission rates, magnetic communications offer unique advantages for applications requiring penetration through dense materials or operation in electromagnetically hostile environments.

Neutrino communication, still in the realm of theoretical research, could eventually provide communication channels that are completely immune to electromagnetic interference. Neutrinos interact so weakly with matter that they can pass through the entire Earth with virtually no attenuation. A neutrino communication system could provide secure, jam-proof communication links between any two points on Earth, or even between Earth and spacecraft anywhere in the solar system.

The technical challenges for neutrino communication are formidable—neutrino detectors currently require massive installations and can detect only a tiny fraction of incident particles. But advances in detector technology and signal processing may eventually make practical neutrino communication systems feasible for critical applications where security and reliability are more important than cost or data rate.

Molecular communication represents another frontier beyond traditional radio. In molecular communication systems, information is encoded in chemical molecules rather than electromagnetic waves. These systems, inspired by biological communication mechanisms, could enable new types of sensor networks that operate at the cellular level within living organisms or in chemical environments where electromagnetic waves cannot propagate.

The Terahertz Frontier

While researchers explore communication paradigms beyond electromagnetic waves, engineers are simultaneously pushing the boundaries of the radio spectrum itself into previously unexplored territory. The terahertz frequency range, between 300 GHz and 3 THz, represents the last major frontier of the electromagnetic spectrum for practical communication applications.

Terahertz frequencies offer enormous bandwidth—potentially thousands of times greater than all lower-frequency radio allocations combined. A single terahertz communication link could theoretically support data rates measured in terabits per second, enabling applications like real-time transmission of ultra-high-definition holographic video or instantaneous backup of entire data centers.

But terahertz communications also present unprecedented technical challenges. At these frequencies, atmospheric absorption is so severe that useful communication ranges are measured in meters rather than kilometers. Water vapor, oxygen, and other atmospheric gases create absorption lines that can completely block terahertz signals over even short distances. Rain, snow, and fog render terahertz communications virtually unusable.

The solution lies in adaptive, short-range networking architectures that use terahertz links for extremely high-bandwidth communication over distances of tens to hundreds of meters, with lower-frequency links providing backup connectivity when atmospheric conditions degrade terahertz performance. Such hybrid networks could enable "terahertz islands" of ultra-high-speed connectivity connected by more conventional wireless backhaul.

Terahertz imaging and sensing applications may prove even more valuable than communications. Terahertz radiation can penetrate many materials that are opaque to visible light while providing much higher resolution than lower-frequency radio waves. This capability enables applications ranging from non-destructive testing of materials to medical imaging systems that can detect cancer cells with unprecedented precision.

The development of practical terahertz systems requires advances in multiple technologies simultaneously. Room-temperature terahertz sources and detectors, currently limited to laboratory demonstrations, must become compact and efficient enough for commercial deployment. Antenna systems must overcome the challenge of coupling efficiently to terahertz radiation while maintaining reasonable physical dimensions. Signal processing algorithms must account for the unique propagation characteristics and noise properties of terahertz systems.

The Internet of Everything

The Internet of Things, already transforming how we think about connected devices, will evolve into what researchers call the "Internet of Everything"—a comprehensive network that connects not just devices and sensors, but materials, environments, and biological systems. This expansion will create unprecedented demands for radio spectrum while simultaneously driving innovations that make more efficient use of available frequencies.

By 2030, industry forecasts predict that over 100 billion connected devices will require some form of wireless connectivity. These devices will range from conventional sensors and actuators to smart materials that can report their own structural health, autonomous vehicles that coordinate with traffic infrastructure, and biological implants that monitor and treat medical conditions in real-time.

The diversity of IoT applications will drive the development of specialized radio technologies optimized for specific use cases. Ultra-low-power radios will enable sensors with decades-long battery life for applications like structural monitoring and environmental sensing.

Ultra-low-latency radios will support real-time control of industrial machinery and autonomous vehicles where millisecond delays could be catastrophic. Ultra-reliable radios will ensure that critical safety systems can communicate even in challenging electromagnetic environments.

The sheer scale of IoT deployments will require new approaches to spectrum management that go far beyond traditional licensing and interference protection models. Machine-to-machine communication protocols will need to automatically negotiate spectrum access among billions of devices without human intervention. AI-powered spectrum management systems will predict and prevent interference before it occurs, automatically reconfiguring networks to maintain performance as deployment patterns change.

The integration of IoT systems with 5G and future cellular networks will create hybrid communication architectures that seamlessly combine licensed and unlicensed spectrum, terrestrial and satellite connectivity, and multiple radio technologies within single devices. A smart city sensor might use ultra-low-power IoT protocols for routine data transmission, automatically switch to cellular networks for urgent alerts, and fall back to satellite communication when terrestrial infrastructure is unavailable.

The Space-Based Infrastructure Revolution

The proliferation of satellite constellations will fundamentally alter the landscape of radio communications by moving much of the world's communication infrastructure from terrestrial towers to orbital platforms. By 2040, mega-constellations containing tens of thousands of satellites will provide global broadband coverage with performance that rivals or exceeds terrestrial fiber optic networks.

This space-based infrastructure will enable communication architectures that are impossible with terrestrial systems. Inter-satellite laser communication links will create a space-based internet backbone that can route traffic between any two points on Earth at nearly the speed of light in vacuum—faster than terrestrial fiber networks that must follow the curved surface of the Earth and deal with the refractive index of glass.

The integration of space-based and terrestrial networks will require new spectrum management approaches that account for the three-dimensional nature of satellite coverage and the rapid movement of satellites across the sky. Dynamic spectrum allocation systems will need to coordinate frequency usage among thousands of satellites while protecting terrestrial services from interference.

Advanced beamforming technologies will enable satellites to create thousands of simultaneous spot beams, each serving different geographic areas with independent frequency reuse. Satellite antennas with hundreds of thousands of elements will focus radio energy with precision measured in meters rather than kilometers, enabling aggressive spatial reuse of spectrum resources.

The economic implications of space-based infrastructure are profound. Once deployed, satellite networks can provide global coverage with marginal costs that approach zero for additional users. This could fundamentally disrupt traditional telecommunications business models based on terrestrial infrastructure investments and geographic service territories.

The Biological Integration

Perhaps the most speculative but potentially transformative development in radio technology is the integration of radio systems with biological organisms. Brain-computer interfaces, already demonstrated in laboratory settings, will evolve to enable direct neural control of radio communication systems. Users will be able to place calls, send messages, and access information through thought alone, without any physical interaction with devices.

The technical challenges for neural radio interfaces are formidable. The electrical signals generated by neurons are extremely weak and variable, requiring sophisticated signal processing to extract meaningful control signals. The radio systems must be biocompatible, reliable over decades of operation, and capable of being powered by biological processes.

But the potential applications extend far beyond convenience. Neural radio interfaces could restore communication abilities to patients with severe disabilities, enable new forms of augmented reality that integrate seamlessly with human perception, and create direct brain-to-brain communication links that transcend the limitations of traditional language.

Biological materials may also find applications in radio system construction. DNA-based data storage systems can achieve information densities that are millions of times greater than conventional electronic storage. Biological sensors can detect chemical and biological agents with sensitivity and specificity that exceed any artificial system. Living materials can self-repair damage and adapt to changing environmental conditions.

The convergence of biotechnology and radio engineering will create hybrid systems that combine the best characteristics of biological and artificial systems. These bio-radio systems could enable new categories of applications ranging from environmental monitoring to medical diagnostics that are impossible with purely artificial technologies.

The End of Spectrum Scarcity

The convergence of multiple technological trends suggests that spectrum scarcity—the fundamental constraint that has shaped radio development since Marconi's first experiments—may eventually become obsolete. This transformation will not come from any single breakthrough but from the cumulative effect of multiple innovations working together.

Cognitive radio technologies will enable automatic, real-time optimization of spectrum usage across millions of devices and frequency bands. AI-powered interference cancellation will allow multiple systems to share the same frequencies without degrading each other's performance. Advanced antenna systems will focus radio energy with precision that enables aggressive spatial reuse of spectrum resources.

The expansion of useful spectrum into terahertz frequencies will provide bandwidth resources that dwarf current allocations. Space-based infrastructure will enable frequency reuse patterns that are impossible with terrestrial systems. Quantum communication systems will provide secure channels that require minimal spectrum resources while offering capabilities that classical systems cannot match.

Perhaps most importantly, the shift from hardware-defined to software-defined radio systems will enable dynamic optimization of spectrum usage based on real-time requirements rather than static allocations determined by regulatory processes. Radios will automatically negotiate spectrum access, share frequencies cooperatively, and adapt their operation to maximize overall system efficiency.

The end of spectrum scarcity will have profound implications for society and technology. The regulatory frameworks that have governed radio for over a century, based on the assumption that spectrum is a scarce resource requiring careful allocation, will need fundamental revision. Business models based on exclusive spectrum access will give way to new approaches based on dynamic resource sharing and optimization.

The Human Element in an Automated World

As radio systems become increasingly automated and intelligent, the role of human operators and engineers will evolve rather than disappear. While AI systems will handle routine tasks like frequency assignment and interference mitigation, humans will remain essential for strategic decision-making, creative problem-solving, and managing the complex social and political aspects of spectrum usage.

The next generation of radio engineers will need skills that combine traditional RF expertise with artificial intelligence, quantum physics, and systems thinking. They will work alongside AI systems as partners rather than being replaced by them, focusing on tasks that require human judgment, creativity, and ethical reasoning.

Amateur radio operators, who have driven many innovations in radio technology since the early 20th century, will continue to play a vital role in exploring new technologies and techniques. Their experimental work will help validate new concepts before they enter commercial deployment, and their emergency communication capabilities will remain essential for disaster response and public safety.

The democratization of radio technology through software-defined systems and AI assistance will enable more people to develop and deploy radio systems without deep technical expertise. This could lead to an explosion of innovation as entrepreneurs and inventors who were previously constrained by the complexity of radio engineering gain access to powerful development tools and deployment platforms.

The Global Coordination Challenge

The technologies that will define radio's future will require unprecedented levels of international cooperation and coordination. Quantum communication systems will need global key distribution networks. AI-powered spectrum management will require real-time coordination across national boundaries. Space-based infrastructure will operate in the global commons of orbital space.

The traditional model of national sovereignty over radio spectrum will need to evolve to accommodate technologies that transcend geographic boundaries. International organizations like the ITU will need new frameworks for managing global radio resources while respecting national interests and security concerns.

The potential for conflict over spectrum resources will increase as more nations and organizations deploy advanced radio systems. Space-based assets will be particularly vulnerable to interference and jamming, creating new categories of international disputes. The development of norms and agreements for responsible behavior in the electromagnetic spectrum will be essential for preventing conflicts that could disrupt global communications.

Environmental and Sustainability Considerations

The future of radio technology must also account for environmental sustainability and energy efficiency. The proliferation of billions of connected devices will create enormous demands for energy, potentially offsetting the environmental benefits of smart grid and environmental monitoring applications.

Advanced radio technologies will need to be designed for minimal environmental impact throughout their lifecycle. This includes using sustainable materials in manufacturing, designing for efficient recycling and disposal, and minimizing energy consumption during operation. The development of energy harvesting technologies that can power radio devices from ambient sources—solar, thermal, vibrational, or electromagnetic energy—will be essential for sustainable IoT deployments.

The electromagnetic pollution created by billions of radio devices will require careful management to prevent interference with sensitive scientific instruments and biological systems. Radio-quiet zones around radio astronomy facilities will need enhanced protection as spectrum usage intensifies. The potential biological effects of long-term exposure to electromagnetic fields will require ongoing research and monitoring.

Looking Beyond 2050

As we project radio technology beyond 2050, the possibilities become increasingly speculative but no less fascinating. The integration of radio systems with advanced materials science could enable programmable matter that can reconfigure its electromagnetic properties on command. Metamaterials with properties that don't exist in nature could enable perfect lenses, invisibility cloaks, and antennas with capabilities that seem to violate physical laws.

The convergence of radio technology with nanotechnology could enable communication systems built at the molecular level, opening possibilities for medical applications that monitor and treat diseases at the cellular level. The integration with quantum mechanics could enable communication protocols based on quantum entanglement that are not limited by the speed of light.

The exploration of space will drive demand for radio systems that can operate over interplanetary and eventually interstellar distances. The challenges of communicating across light-years will require technologies that push the boundaries of what is physically possible, potentially leading to discoveries that revolutionize our understanding of the universe itself.

The Continuing Journey

As we reach the end of our journey through the history and future of radio, it's worth reflecting on how far we've come since Marconi first transmitted signals across the Atlantic in 1901. The simple spark-gap transmitter and coherer detector that enabled that historic achievement have evolved into a global ecosystem of unprecedented complexity and capability.

Yet the fundamental principle remains unchanged: radio waves carry information by modulating the electromagnetic field, and receivers extract that information by detecting the modulation. The mathematical equations that Maxwell wrote in the 19th century still govern how radio waves propagate through space. The physical limits imposed by the speed of light and the finite nature of the electromagnetic spectrum still constrain what is possible.

What has changed is our understanding of how to work within these constraints, how to extract maximum utility from the finite resource of radio spectrum, and how to create systems of incredible sophistication that seem to approach the theoretical limits of what physics allows.

The technologies we've explored in this final chapter—artificial intelligence, quantum communications, terahertz systems, space-based infrastructure, and biological integration—represent not an end point but waypoints on a continuing journey of discovery and innovation. Each breakthrough opens new possibilities while revealing new challenges. Each solution creates new problems to solve.

The invisible waves that carry our radio signals will continue to connect humanity across distances both terrestrial and cosmic. They will carry our voices, our dreams, and our most important communications as we explore the solar system and eventually venture to the stars. The radio spectrum that seemed infinite to early pioneers but finite to modern engineers may yet prove to have capabilities we haven't imagined.

The story of radio is ultimately a human story—a tale of ingenuity, creativity, and the relentless drive to communicate across distance and time. As we face the challenges and opportunities of the decades ahead, we carry with us the accumulated wisdom of more than a century of radio innovation and the unshakeable belief that human creativity can overcome any obstacle.

The horizon ahead is bright with possibility. The invisible waves that have carried us this far will carry us forward into a future where the boundaries between the possible and impossible continue to blur, where the spectrum that once seemed so limited reveals new dimensions of capability, and where radio technology continues to transform how we understand and interact with our world.

In the end, radio technology is not just about transmitting information—it's about connecting humanity, enabling discovery, and expanding the boundaries of what we can achieve together. As we stand on the threshold of that future, we can hear in the invisible waves around us the echo of all the voices that have come before and the promise of all the communications yet to come.

The final frontier of radio is not a destination but a direction—always beyond the horizon, always beckoning us forward, always revealing new wonders in the invisible ocean of electromagnetic waves that surrounds us all. The journey continues, and the best is yet to come.

Appendices

Appendix A

Major International Radio Conferences and Treaties

This appendix provides a chronological reference to the most significant international conferences and treaties that have shaped radio communication policy and spectrum allocation throughout history. These gatherings represent the collective effort of the global community to coordinate the use of the electromagnetic spectrum and ensure interference-free operation of radio services worldwide.

Early Maritime Safety Era (1903-1912)

Berlin Preliminary Conference on Wireless Telegraphy (1903)

The first international attempt to coordinate wireless telegraphy regulations, attended by nine European nations. While no binding agreements were reached, this conference established the precedent for international cooperation in radio matters.

First International Radiotelegraph Conference - Berlin (1906)

Participants: 30 maritime nations

Key Achievements:

- Established the first International Radiotelegraph Convention
- Created uniform regulations for ship wireless stations
- Mandated that coastal stations communicate with all ships regardless of wireless equipment manufacturer
- Established international distress frequencies
- Required ships to maintain 24-hour radio watch

This conference marked ITU's entry into radio communications and established the fundamental principle that radio waves, crossing all national boundaries, required international coordination.

London Radiotelegraph Conference (1912)

Participants: 45 nations

Key Achievements:

- Revised the Berlin Convention in response to the Titanic disaster
- Strengthened requirements for ship radio equipment and operators
- Established standardized distress procedures and frequencies
- Created the SOS distress signal as the international standard
- Mandated continuous radio watch on passenger vessels

The Interwar Period (1927-1938)

International Radiotelegraph Conference - Washington (1927)

Key Achievements:

- Established the International Radio Consultative Committee (CCIR) to study technical and operating questions
- Created systematic frequency allocation plans
- Recognized broadcasting as a legitimate radio service requiring protection
- Established technical standards for international radio services
- Divided the world into three regions for frequency planning purposes

International Telecommunication Conference - Madrid (1932)

Key Achievements:

- Merged telegraph and radiotelegraph regulations into single International Telecommunication Convention
- Established the International Telecommunication Union (ITU) as successor to previous organizations
- Created comprehensive frequency allocation table
- Recognized television broadcasting as a radio service
- Established notification procedures for frequency assignments

World War II and Reconstruction (1938-1959)

International Telecommunication Conference and International Radio Conference - Atlantic City (1947)

Key Achievements:

- Created the International Frequency Registration Board (IFRB) to regulate frequency use
- Established post-war frequency allocation plans
- Recognized FM broadcasting and television services
- Created procedures for registering frequency assignments in the International Frequency List
- Established interference resolution procedures

Administrative Radio Conference - Geneva (1949)

Key Achievements:

- Updated frequency allocations for Regions 1 and 3
- Addressed post-war reconstruction needs for European and Asian radio services
- Established technical standards for VHF and UHF services
- Created framework for international broadcasting coordination

Administrative Radio Conference - Geneva (1959)

Key Achievements:

- Comprehensive revision of Radio Regulations
- Established frequency plans for space services
- Created technical standards for stereo FM broadcasting
- Updated maritime and aeronautical radio procedures
- Established framework for satellite coordination

The Space Age (1963-1979)

Extraordinary Administrative Radio Conference (Space) - Geneva (1963)

Participants: ITU member countries

Key Achievements:

- First conference dedicated entirely to space communications
- Allocated specific frequency bands for space services
- Established technical criteria for satellite systems
- Created coordination procedures for satellite networks

- Set foundation for all subsequent space communication regulations

Space Radio Conference - Geneva (1971)

Key Achievements:

- Expanded allocations for satellite broadcasting services
- Established technical standards for satellite earth stations
- Created procedures for geostationary satellite coordination
- Allocated frequencies for satellite navigation services

World Administrative Radio Conference - Geneva (1979)

Key Achievements:

- Comprehensive revision of Radio Regulations
- Updated frequency allocations for all services
- Established detailed procedures for satellite coordination
- Created framework for cellular mobile services
- Allocated spectrum for emerging digital technologies

The Cellular and Digital Era (1987-1995)

Mobile Radio Conference - Geneva (1987)

Key Achievements:

- Allocated frequencies for cellular mobile services
- Established technical standards for digital mobile systems
- Created framework for international roaming
- Allocated spectrum for satellite mobile services

World Administrative Radio Conference - Málaga-Torremolinos (1992)

Key Achievements:

- Updated mobile service allocations
- Established framework for personal communications services
- Created procedures for high-frequency coordination
- Allocated spectrum for mobile satellite services

World Radiocommunication Conference - Geneva (1995)

- First conference under reformed ITU structure
- Comprehensive update of Radio Regulations
- Established framework for digital broadcasting services
- Updated satellite service procedures
- Created allocations for emerging wireless technologies

The Internet and Convergence Era (1997-2015)

World Radiocommunication Conference 1997 (WRC-97) - Geneva

Key Achievements:

- Allocated spectrum for third-generation mobile systems (3G)
- Established framework for satellite personal communications
- Updated broadcasting service allocations
- Created procedures for digital audio broadcasting

World Radiocommunication Conference 2000 (WRC-00) - Istanbul

Key Achievements:

- Finalized 3G mobile spectrum allocations globally
- Established International Mobile Telecommunications-2000 (IMT-2000) framework
- Updated satellite service regulations
- Created allocations for satellite digital audio radio

World Radiocommunication Conference 2007 (WRC-07) - Geneva

Key Achievements:

- Allocated spectrum for International Mobile Telecommunications-Advanced (IMT-Advanced/4G)
- Updated aeronautical mobile satellite service allocations
- Established framework for broadcasting-satellite service
- Created procedures for coordinating terrestrial and satellite services

World Radiocommunication Conference 2012 (WRC-12) - Geneva

Key Achievements:

- Expanded mobile broadband allocations
- Updated satellite service procedures
- Established framework for cognitive radio systems
- Created allocations for radio astronomy protection

World Radiocommunication Conference 2015 (WRC-15) - Geneva

Key Achievements:

- Allocated additional spectrum for International Mobile Telecommunications (IMT/4G and future 5G)
- Updated satellite service allocations
- Established procedures for wireless broadband in aviation
- Created framework for Internet of Things applications

The 5G and IoT Era (2019-Present)

World Radiocommunication Conference 2019 (WRC-19) - Sharm El Sheikh

Key Achievements:

- Identified approximately 17 GHz of spectrum for International Mobile Telecommunications (5G)
- Updated satellite service procedures for mega-constellations
- Established framework for high-altitude platform systems
- Created allocations for railway radiocommunication systems
- Updated procedures for cognitive radio systems

World Radiocommunication Conference 2023 (WRC-23) - Dubai

Key Achievements:

- Further spectrum identification for 5G and future 6G systems
- Updated regulations for satellite constellations
- Established framework for wireless power transmission
- Created procedures for transportation and safety services
- Updated maritime and aeronautical mobile service regulations

Regional Conferences

European Conference of Postal and Telecommunications Administrations (CEPT)

Key Regional Achievements:

- Coordinated GSM frequency allocations across Europe
- Established unified approach to spectrum harmonization
- Created framework for cross-border coordination
- Developed technical standards for regional services

Inter-American Telecommunication Commission (CITEL)

Key Regional Achievements:

- Coordinated spectrum use across the Americas
- Established regional frequency plans
- Created procedures for cross-border coordination
- Developed framework for regional mobile services

Asia-Pacific Telecommunity (APT)

Key Regional Achievements:

- Harmonized mobile spectrum across Asia-Pacific region
- Coordinated satellite service procedures
- Established regional framework for emerging technologies
- Created cross-border coordination procedures

Key Treaties and Agreements

International Telecommunication Constitution and Convention

The fundamental treaty governing international telecommunications, periodically updated at Plenipotentiary Conferences. Establishes ITU structure, basic principles, and member obligations.

Radio Regulations

The international treaty governing use of radio spectrum, updated at each World Radiocommunication Conference. Contains detailed frequency allocations, technical standards, and operational procedures.

Regional Agreements

Various regional agreements supplement global regulations:

- European Broadcasting Area Agreement (Geneva, 1984)
- Regional Agreement for Broadcasting Satellite Service (Geneva, 1977)
- Regional Agreement for Aeronautical Mobile Service (Geneva, 1990)

Specialized Conferences

Broadcasting Conferences

- Administrative Radio Conference for Broadcasting (Geneva, 1984)
- Regional Administrative Conference for Digital Audio Broadcasting (Geneva, 1995)
- Conference on Digital Terrestrial Broadcasting (Chester, 1997)

Maritime and Aeronautical Conferences

- World Administrative Radio Conference for Maritime Mobile Service (Geneva, 1967)
- World Administrative Radio Conference for Aeronautical Mobile Service (Geneva, 1978)
- Global Maritime Distress and Safety System Conference (London, 1988)

Satellite Coordination Meetings

Regular coordination meetings under ITU procedures for specific satellite services and orbital positions, ensuring interference-free operation of satellite networks worldwide.

Looking Forward: Planned Conferences

World Radiocommunication Conference 2027 (WRC-27)

Anticipated Agenda Items:

- Spectrum requirements for 6G and beyond
- Regulatory framework for quantum communications
- Updated procedures for mega-constellation coordination
- Framework for space-based infrastructure services
- Allocation considerations for terahertz communications

This chronological record demonstrates the evolution of international radio cooperation from basic maritime safety concerns to the complex, technology-driven spectrum management challenges of the modern era. Each conference has built upon previous work while adapting to new technologies and changing global communication needs, creating the regulatory framework that enables today's interconnected world.

The ongoing success of international radio coordination through these conferences and treaties represents one of humanity's most successful examples of technical cooperation across national boundaries, ensuring that the invisible waves carrying our communications continue to serve all nations and peoples effectively.

Appendix B

Amateur Radio Band Plans by Region

Amateur radio band plans represent one of the most successful examples of international spectrum coordination in radio history. Unlike commercial services that rely heavily on regulatory enforcement, amateur radio band plans operate primarily through voluntary cooperation among radio amateurs worldwide. This appendix provides a comprehensive reference to amateur radio band plans across the three International Amateur Radio Union (IARU) regions, reflecting both historical development and current practice as of 2025.

Introduction to Amateur Radio Band Planning

The International Amateur Radio Union divides the world into three regions for the purpose of frequency coordination and band planning:

- Region 1: Europe, Africa, the Middle East, former Soviet Union, and Mongolia
- Region 2: The Americas (North, Central, and South America)
- Region 3: Asia-Pacific (excluding Region 1 countries), Australia, and New Zealand

Each regional organization develops band plans to offer guidance to radio amateurs on how different operating interests and modes of emission can be best accommodated within the limited spectrum available to the amateur services. Observing band plans is considered good amateur operating practice, and in most cases adherence is voluntary, although some administrations incorporate regional band plans into national regulations.

The Philosophy of Amateur Band Planning

Amateur radio band planning differs fundamentally from commercial spectrum management in several key ways:

- Voluntary Compliance: Band plans rely on the self-discipline and courtesy of amateur operators rather than regulatory enforcement
- Technical Compatibility: Plans are designed to group compatible operating modes and separate incompatible ones
- Shared Use: Most amateur frequency allocations accommodate multiple users simultaneously through various sharing techniques
- Innovation Accommodation: Band plans must balance established uses with space for experimental and emerging technologies

IARU Region 1 Band Plans

Region 1, covering Europe, Africa, and parts of Asia, maintains the most detailed and widely adopted amateur radio band plans. The European Conference of Postal and Telecommunications Administrations (CEPT) has been instrumental in harmonizing amateur allocations across national boundaries.

HF Band Plans (Region 1)

160 Meters (1.810-2.000 MHz)

Frequency (MHz)	Usage	Notes
1.810-1.838	CW, narrow-band digital modes	QRP center 1.836
1.838-1.840	Narrow-band digital modes	
1.840-1.843	All modes	Digital voice center 1.842
1.843-2.000	All modes	Regional emergency frequencies

80 Meters (3.500-3.800 MHz)

Frequency (MHz)	Usage	Notes
3.500-3.510	CW, narrow-band digital modes	
3.510-3.560	CW	Contest preferred
3.560-3.580	CW, narrow-band digital modes	
3.580-3.590	Narrow-band digital modes	
3.590-3.600	Narrow-band digital modes	Digital voice
3.600-3.650	All modes (SSB)	Regional emergency
3.650-3.700	All modes (SSB)	
3.700-3.800	All modes (SSB)	Contest preferred

40 Meters (7.000-7.200 MHz)

Frequency (MHz)	Usage	Notes
7.000-7.040	CW	
7.040-7.047	CW, narrow-band digital modes	
7.047-7.050	Narrow-band digital modes	
7.050-7.060	All modes (SSB)	
7.060-7.100	All modes (SSB)	Emergency traffic
7.100-7.130	All modes (SSB)	
7.130-7.175	All modes (SSB)	Contest preferred
7.175-7.200	All modes (SSB)	

VHF/UHF Band Plans (Region 1)

2 Meters (144-146 MHz)

Recent updates in 2025 have incorporated additional FM gateway frequencies to accommodate increased activity following new licensing initiatives.

Frequency (MHz)	Usage	Notes
144.000-144.025	Amateur-satellite service	Uplink
144.025-144.150	CW, SSB	EME activity center 144.050

144.150-144.180	CW, SSB	General usage
144.180-144.300	Packet radio, digital modes	
144.300-144.500	Digital modes, beacons	
144.500-144.794	FM simplex	
144.794-144.990	FM repeater outputs	
145.000-145.194	FM repeater outputs	
145.194-145.206	FM gateways	Enhanced in 2025
145.206-145.594	FM repeater outputs	
145.594-145.794	FM simplex	
145.794-146.000	FM repeater inputs	

70 Centimeters (430-440 MHz)

The 2025 band plan incorporated significant changes to narrowband sections, with unified 2.7 kHz maximum bandwidth for all modes in line with IARU Region 1 recommendations.

Frequency (MHz)	Usage	Notes
430.000-431.000	All modes, narrowband	2.7 kHz max BW (2025 update)
431.000-432.000	All modes, narrowband	
432.000-432.025	EME	
432.025-432.150	CW, SSB	

| 432.150-432.400 | Digital modes | |

| 432.400-432.500 | Beacons | |

| 432.500-433.000 | All modes | |

| 433.000-434.594 | Digital modes | |

| 434.594-434.981 | FM repeater outputs | |

| 435.000-438.000 | Amateur-satellite service | |

| 438.000-440.000 | FM, digital | Various splits including new 5 MHz and 9 MHz emergency splits |

IARU Region 2 Band Plans

Region 2, covering the Americas, operates under band plans coordinated through the Inter-American Telecommunication Commission (CITEL) and implemented by national amateur organizations such as the ARRL in the United States.

US Amateur Band Plans (ARRL)

The ARRL band plans serve as the de facto standard for Region 2, with variations adopted by other countries in the region based on their specific frequency allocations and local needs.

10 Meters (28.0-29.7 MHz)

| Frequency (MHz) | Usage | License Class | Notes |

|----------------|-------|--------------|--------|
28.000-28.070	CW	All classes	
28.070-28.150	RTTY/Data	All classes	
28.150-28.190	CW	All classes	
28.200-28.300	Beacons	All classes	
28.300-29.000	Phone	All classes	
29.000-29.200	AM/FM	All classes	
29.200-29.300	Fast-scan TV	All classes	
29.300-29.510	Satellite downlinks	All classes	
29.510-29.590	Repeater inputs	All classes	
29.590-29.610	FM simplex	All classes	
29.610-29.700	Repeater outputs	All classes	

1.25 Meters (222-225 MHz)

This band exemplifies the detailed coordination required for VHF/UHF band planning:

Frequency (MHz)	Usage	Notes
222.000-222.025	EME	Earth-Moon-Earth communication
222.025-222.050	Weak signal	
222.050-222.060	Propagation beacons	
222.100	SSB/CW calling frequency	

222.100-222.150	Weak signal CW & SSB	
222.150-222.250	Local coordinator option	
222.250-223.380	FM repeater inputs	25 kHz spacing
223.400-223.520	FM simplex	
223.520-223.640	Digital, packet	
223.640-223.700	Links, control	
223.710-223.850	Local coordinator option	
223.850-224.980	Repeater outputs	

23 Centimeters (1240-1300 MHz)

This microwave band demonstrates the complexity of modern amateur band planning:

Frequency (MHz)	Usage	Notes
1240-1246	ATV Channel 1	
1246-1248	FM/Digital links	Paired with 1258-1260
1248-1252	Digital	
1252-1258	ATV Channel 2	
1258-1260	FM/Digital links	Paired with 1246-1248
1260-1270	Satellite uplinks, experimental	
1270-1276	FM/Digital repeater inputs	25 kHz spacing
1276-1282	ATV Channel 3	

| 1282-1288 | FM/Digital repeater outputs | |

| 1288-1294 | Experimental, simplex ATV | |

| 1294-1295 | FM simplex | 1294.500 calling frequency |

| 1295.000-1297.000 | Narrowband segment | |

| 1295.800-1296.080 | EME | |

| 1296.080-1296.200 | Weak signal | 1296.100 calling frequency |

| 1296.200-1296.400 | Beacons | |

| 1296.400-1297.000 | General narrowband | |

License Class Considerations

Region 2 band plans must accommodate different license classes with varying privileges. The Amateur Extra class enjoys the broadest access across all amateur bands, while Technician and General class licensees have more limited HF privileges but full access to VHF/UHF bands.

IARU Region 3 Band Plans

Region 3, covering Asia-Pacific, operates under band plans coordinated through the Asia-Pacific Telecommunity (APT) with significant input from national organizations such as the Japanese Amateur Radio League (JARL) and Wireless Institute of Australia (WIA).

Key Features of Region 3 Band Plans

- Harmonization with ITU Region 3 allocations: Band plans align with ITU frequency allocations specific to the Asia-Pacific region
- Accommodation of high population density: Plans account for intense amateur activity in countries like Japan
- Emergency communication provisions: Special emphasis on disaster communication frequencies
- Satellite coordination: Extensive provisions for amateur satellite operations

HF Variations in Region 3

Region 3 amateur allocations differ from other regions in several key bands:

40 Meters

- Region 3 allocation: 7.000-7.300 MHz (compared to 7.000-7.200 MHz in Region 1)

- Enhanced phone segment: 7.100-7.300 MHz

- Digital modes: 7.035-7.045 MHz

20 Meters

- Consistent with other regions: 14.000-14.350 MHz

- Strong DX activity due to Pacific propagation paths

- Emergency frequency coordination with Pacific island nations

Band Plan Evolution and Recent Changes

Amateur radio band plans continue to evolve to accommodate new technologies and operating practices. The 2025 updates highlight several ongoing trends:

Digital Mode Integration

Modern band plans increasingly accommodate digital modes that were unknown when original plans were developed. Recent changes include:

- Dedicated subbands for digital voice modes
- Provisions for weak-signal digital protocols
- Integration of internet-linked repeater systems

Emergency Communication Evolution

The 2025 Region 1 updates include new emergency communication provisions, including 9 MHz split repeaters specifically for emergency and community usage. This reflects the growing importance of amateur radio in disaster communication.

Microwave Band Development

Ongoing work in microwave bands reflects the expansion of amateur activity into millimeter-wave frequencies. The RSGB submitted options to the 2025 IARU Interim Meeting regarding potential changes to the 23 cm band.

IoT and Digital Integration

Modern band plans must accommodate the integration of amateur radio with Internet of Things (IoT) technologies, including provisions for LoRa and other low-power digital protocols.

Coordination Challenges and Solutions

Cross-Border Coordination

Amateur radio band plans face unique challenges due to the international nature of amateur radio propagation:

- HF propagation: Signals routinely cross multiple countries and regions
- VHF/UHF enhancement: Sporadic E and other propagation modes create unexpected long-distance paths
- Satellite footprints: Amateur satellites serve multiple regions simultaneously

Interference Mitigation

Band plans employ several techniques to minimize interference:

- Mode segregation: Grouping compatible modes and separating incompatible ones
- Power level coordination: Different power levels for different activities
- Time-based coordination: Contest periods and special event coordination

- Geographic coordination: Regional variations within global frameworks

Technology Accommodation

Modern band plans must balance established uses with emerging technologies:

- Legacy system protection: Maintaining space for established modes
- Innovation encouragement: Providing experimental space for new technologies
- Standards development: Coordinating technical standards for new modes
- Transition management: Gradual migration between old and new technologies

Special Considerations

Contest and DXpedition Coordination

Band plans include provisions for contest activities and DXpeditions that generate intense activity:

- Contest-preferred frequency segments
- DX calling frequencies and procedures
- Split-frequency operation guidelines
- Pile-up management techniques

Beacon and Propagation Research

Amateur radio band plans provide for propagation research through:

- Dedicated beacon subbands
- Coordinated beacon frequencies for propagation study
- Weak-signal research segments
- Time-based beacon coordination

Emergency and Public Service

Band plans increasingly emphasize emergency communication:

- Dedicated emergency frequencies
- ARES/RACES coordination
- Health and welfare traffic handling
- Interoperability with public safety services

Future Trends in Amateur Band Planning

As amateur radio continues to evolve, several trends are shaping future band plan development:

Software-Defined Radio (SDR) Integration

The increasing use of SDR technology enables more flexible band plan implementation and real-time coordination between incompatible modes.

Artificial Intelligence Applications

AI-powered band planning systems may eventually optimize frequency usage in real-time based on propagation conditions, activity levels, and interference patterns.

Millimeter-Wave Expansion

Growing amateur interest in millimeter-wave frequencies requires new approaches to band planning that account for unique propagation characteristics and technical requirements.

Internet Integration

The integration of internet linking with traditional amateur radio requires new band planning approaches that accommodate hybrid systems.

Conclusion

Amateur radio band plans represent a remarkable achievement in international cooperation and technical coordination. Through voluntary compliance and mutual respect, amateur radio operators worldwide have created a framework that enables millions of radio amateurs to share finite spectrum resources while pursuing diverse technical and operational interests.

The continued evolution of these band plans, as demonstrated by the 2025 updates, shows that amateur radio remains a dynamic and innovative service that adapts to new technologies while maintaining its fundamental principles of self-regulation and technical excellence. As we look toward the future, amateur radio band plans will continue to serve as a model for how technical communities can successfully manage shared resources through cooperation rather than regulation.

The success of amateur radio band planning offers lessons for spectrum management in other services, demonstrating that voluntary coordination based on technical merit and mutual respect can achieve results that exceed what regulatory enforcement alone might accomplish. This legacy of cooperation will be essential as the radio spectrum becomes increasingly crowded and the need for efficient, interference-free operation becomes ever more critical.

Appendix C

Timeline of Major Radio Astronomy Discoveries

This appendix chronicles the major discoveries and milestones in radio astronomy from its theoretical foundations in the 19th century through the present day. Radio astronomy has revealed a universe invisible to optical telescopes, uncovering phenomena from pulsars to cosmic background radiation that have revolutionized our understanding of the cosmos.

Pre-Discovery Era (1860s-1920s)

1864

James Clerk Maxwell publishes "A Dynamical Theory of the Electromagnetic Field," establishing that electromagnetic radiation can exist at any wavelength, laying the theoretical foundation for radio astronomy.

1886-1888

Heinrich Hertz demonstrates the existence of radio waves in laboratory experiments, confirming Maxwell's theoretical predictions.

1896

German astrophysicists Johannes Wilsing and Julius Scheiner attempt the first systematic search for radio emission from the Sun. Technical limitations prevent detection.

1897-1900

Oliver Lodge constructs a centimeter-wave radiation apparatus in an attempt to detect solar radio emission. The experiment fails due to instrument sensitivity limitations.

1902

Discovery of the ionosphere leads physicists to conclude erroneously that this radio-reflecting layer would prevent detection of astronomical radio sources from Earth's surface.

The Birth of Radio Astronomy (1930s)

1930-1932

Karl Jansky, a radio engineer at Bell Telephone Laboratories, discovers the first astronomical radio source while investigating static interference with transatlantic voice transmissions. His directional antenna detects a persistent "hiss" that repeats every 23 hours and 56 minutes—exactly one sidereal day.

1932

Jansky identifies the radio source as the center of the Milky Way galaxy in the constellation Sagittarius, marking the birth of radio astronomy.

1933

Jansky publishes his findings, but the astronomical community initially shows little interest in this new form of observation.

1937-1938

Grote Reber, an amateur radio operator and engineer, builds the first purpose-built radio telescope in his backyard in Wheaton, Illinois—a 31-foot parabolic dish operating at 160 MHz.

1939

Reber confirms Jansky's discovery and creates the first radio map of the sky, detecting radio emission from the Milky Way and other sources.

World War II and Post-War Development (1940s-1950s)

1942

British radar operators detect intense radio noise from the Sun, leading to the first solar radio astronomy observations during World War II.

1945

End of World War II releases radar technology and expertise for peaceful astronomical research, accelerating radio astronomy development.

1946

First radar echoes are bounced off the Moon by the U.S. Army Signal Corps, demonstrating radar astronomy techniques.

1948

Australian radio astronomers led by John Bolton begin systematic surveys of discrete radio sources in the southern hemisphere.

1949

Bolton identifies the first radio galaxy, Centaurus A, associating it with the optical galaxy NGC 5128.

1950

Harvard physicists predict the 21-centimeter hydrogen line, the most important spectral line in radio astronomy.

1951

Harold "Doc" Ewen and Edward Purcell detect the 21-centimeter hydrogen line using horn antenna equipment at Harvard University, opening the field of radio spectroscopy.

1951

Dutch astronomers Hendrik van de Hulst, C. Alex Muller, and Jan Oort use the 21-cm line to map the spiral structure of the Milky Way galaxy.

1955

Bernard Burke and Kenneth Franklin at the Carnegie Institution detect radio bursts from Jupiter, discovering the first planetary radio source.

1957

The 250-foot Jodrell Bank radio telescope becomes operational in England under Sir Bernard Lovell's direction.

The Space Age and Major Discoveries (1960s)

1958

Frank Drake and Harold Ewen detect radio emission from Saturn and a planetary nebula at 8000 MHz.

1958

First radar echoes from Venus are obtained at Lincoln Laboratory.

1958

The first MASER (Microwave Amplification by Stimulated Emission of Radiation) is used in radio astronomy, dramatically improving receiver sensitivity.

1959

First trans-Atlantic communication is achieved via lunar reflection, demonstrating radar astronomy capabilities.

1960

Frank Drake conducts Project Ozma, the first scientific search for extraterrestrial intelligence (SETI), using the 85-foot telescope at Green Bank.

1962

Cyrus Hazard, Antony Hewish, and others use lunar occultation to precisely locate the radio source 3C 273, leading to the discovery of quasars.

1963

Maarten Schmidt identifies 3C 273 as the first quasar, revolutionizing understanding of active galactic nuclei and cosmology.

1965

Arno Penzias and Robert Wilson at Bell Labs discover the cosmic microwave background radiation at 7.35 cm wavelength, providing crucial evidence for the Big Bang theory.

The Pulsar Revolution (Late 1960s)

1967

Jocelyn Bell, working with Antony Hewish at Cambridge University, discovers the first pulsar (CP 1919) using a radio telescope designed to study scintillation.

1968

Large, Vaughan, and Mills discover the Vela Pulsar via radio observations.

1968

Frank Lovelace discovers the Crab Pulsar using the Arecibo Radio Telescope, linking pulsars to supernova remnants.

1968

Thomas Gold proposes that pulsars are rapidly rotating neutron stars, explaining their regular radio pulses.

Advanced Techniques and New Discoveries (1970s)

1970

The first Very Long Baseline Interferometry (VLBI) observations achieve angular resolutions better than optical telescopes.

1973

Bruce Turner and Meudon Observatory observers simultaneously discover OH absorption in Comet Kohoutek, marking the first confirmed comet detection at radio wavelengths.

1973

Robert Brown and Morton Roberts discover a narrow radio absorption line in the quasar 3C 286, identifying it as a redshifted 21-cm line at z = 0.692.

1974

Russell Hulse and Joseph Taylor discover the first binary pulsar (PSR B1913+16), later providing evidence for gravitational waves.

1975

The Arecibo Observatory transmits the first intentional radio message to potential extraterrestrial civilizations.

The Era of Large Arrays (1980s-1990s)

1980

The Very Large Array (VLA) in New Mexico becomes fully operational with 27 antennas, providing unprecedented resolution and sensitivity.

1982

First detection of water masers in external galaxies, opening the field of extragalactic maser astronomy.

1988

Discovery of the first millisecond pulsar, PSR J1937+21, spinning 642 times per second.

1990

Launch of the Cosmic Background Explorer (COBE) satellite provides detailed measurements of the cosmic microwave background.

1993

Hulse and Taylor receive the Nobel Prize in Physics for their discovery of the binary pulsar and indirect confirmation of gravitational waves.

1995

Discovery of the first exoplanet around a pulsar (PSR B1257+12) using precise radio timing measurements.

The Digital Revolution (2000s-2010s)

2003

The Spitzer Space Telescope begins operations, complementing radio observations with infrared data.

2007

Discovery of Fast Radio Bursts (FRBs) by Duncan Lorimer and his team, revealing a new class of cosmic phenomena.

2015

The Laser Interferometer Gravitational-Wave Observatory (LIGO) directly detects gravitational waves, confirming predictions based on radio pulsar observations.

2016

Breakthrough Listen, the most comprehensive SETI program to date, begins systematic radio surveys for technosignatures.

2017

Event Horizon Telescope (EHT) collaboration begins observations that will lead to the first image of a black hole shadow.

Modern Era and Future Prospects (2020s-Present)

2019

Event Horizon Telescope releases the first image of a black hole (M87), combining radio observations from telescopes worldwide.

2020

Canadian Hydrogen Intensity Mapping Experiment (CHIME) detects periodic fast radio bursts, suggesting some FRBs originate from magnetars.

2021

Ongoing construction of the Square Kilometre Array (SKA) begins, promising to revolutionize radio astronomy with unprecedented sensitivity and resolution.

2022

James Webb Space Telescope begins operations, providing complementary infrared observations to radio astronomy discoveries.

2023

Continued discoveries of repeating FRBs and their host galaxies advance understanding of these mysterious phenomena.

2024

Advanced VLBI techniques achieve microarcsecond resolution, enabling detailed studies of black hole environments and stellar formation regions.

2025

Current developments include enhanced AI-assisted data analysis, improved interference mitigation techniques, and preparations for next-generation radio telescope arrays.

Key Technological Milestones

Antenna and Telescope Development

- 1937: First purpose-built radio telescope (Grote Reber)
- 1957: Jodrell Bank 250-foot telescope
- 1963: Arecibo 1000-foot dish
- 1980: VLA becomes operational
- 2021: SKA construction begins

Receiver Technology

- 1958: First astronomical use of masers
- 1965: Development of parametric amplifiers
- 1980s: Introduction of superconducting receivers
- 2000s: Digital signal processing revolution
- 2020s: AI-enhanced signal detection

Observational Techniques

- 1946: First radar astronomy (lunar echoes)
- 1951: Radio spectroscopy (21-cm line)
- 1962: Lunar occultation techniques
- 1970: Very Long Baseline Interferometry
- 2019: Event Horizon Telescope imaging

Impact on Fundamental Science

Radio astronomy has fundamentally transformed our understanding of the universe:

Cosmology

- Cosmic microwave background discovery confirmed Big Bang theory
- Quasar discoveries revealed active galactic nuclei
- Radio galaxy surveys mapped large-scale structure

Stellar Evolution

- Pulsar discoveries revealed neutron stars
- Binary pulsar observations confirmed general relativity
- Radio observations of stellar formation regions

Planetary Science

- Discovery of planetary magnetospheres
- Radar mapping of planetary surfaces
- Detection of organic molecules in space

Astrobiology

- Search for extraterrestrial intelligence
- Detection of complex organic molecules
- Study of potentially habitable exoplanets

Regional and International Contributions

United States

- Bell Labs discoveries (cosmic background, pulsars)
- National Radio Astronomy Observatory facilities
- Arecibo Observatory contributions

United Kingdom

- Cambridge pulsar discoveries
- Jodrell Bank pioneering work
- MERLIN interferometer network

#Australia

- Southern hemisphere surveys
- Australia Telescope National Facility
- Square Kilometre Array development

Europe

- European Southern Observatory
- Very Long Baseline Interferometry networks
- Low-frequency radio astronomy

International Collaboration

- Event Horizon Telescope global cooperation
- Square Kilometre Array partnership
- International VLBI networks

Future Horizons

The timeline of radio astronomy continues with ambitious projects and technological developments:

- Square Kilometre Array: Will provide 50 times the sensitivity of current instruments
- Lunar radio telescopes: Far-side Moon observations free from terrestrial interference
- Space-based VLBI: Extending baselines beyond Earth's surface
- AI and machine learning: Automated discovery of new phenomena
- Multi-messenger astronomy: Combining radio with gravitational wave and neutrino observations

This timeline demonstrates radio astronomy's evolution from accidental discovery to one of the most important tools for understanding the universe. Each decade has brought revolutionary discoveries that have reshaped our cosmic perspective, and the field continues to push the boundaries of technology and knowledge as we explore the radio universe with ever-greater sensitivity and resolution.

The invisible waves that carry radio signals from the cosmos continue to reveal secrets of the universe, from the birth of stars to the nature of space and time itself. Radio astronomy remains at the forefront of astronomical discovery, promising continued revelations about our place in the cosmic story.

Appendix D

Frequency Allocation Chart (3 kHz to 300 GHz)

United States Frequency Allocation – Radio Spectrum:

https://www.ntia.gov/files/ntia/publications/2003-allochrt.pdf

UNITED STATES FREQUENCY ALLOCATIONS

THE RADIO SPECTRUM

RADIO SERVICES COLOR LEGEND

AERONAUTICAL MOBILE	INTER-SATELLITE	RADIO ASTRONOMY
AERONAUTICAL MOBILE SATELLITE	LAND MOBILE	RADIODETERMINATION SATELLITE
AERONAUTICAL RADIONAVIGATION	LAND MOBILE SATELLITE	RADIOLOCATION
AMATEUR	MARITIME MOBILE	RADIOLOCATION SATELLITE
AMATEUR SATELLITE	MARITIME MOBILE SATELLITE	RADIONAVIGATION
BROADCASTING	MARITIME RADIONAVIGATION	RADIONAVIGATION SATELLITE
BROADCASTING SATELLITE	METEOROLOGICAL	SPACE OPERATION
EARTH EXPLORATION SATELLITE	METEOROLOGICAL SATELLITE	SPACE RESEARCH
FIXED	MOBILE	STANDARD FREQUENCY AND TIME SIGNAL
FIXED SATELLITE	MOBILE SATELLITE	STANDARD FREQUENCY AND TIME SIGNAL SATELLITE

ACTIVITY CODE

GOVERNMENT EXCLUSIVE

GOVERNMENT/NON-GOVERNMENT SHARED

NON-GOVERNMENT EXCLUSIVE

ALLOCATION USAGE DESIGNATION

SERVICE	EXAMPLE	DESCRIPTION
Primary	FIXED	Capital Letters
Secondary	Mobile	1st Capital with lower case letters

U.S. DEPARTMENT OF COMMERCE
National Telecommunications and Information Administration
Office of Spectrum Management
August 2011

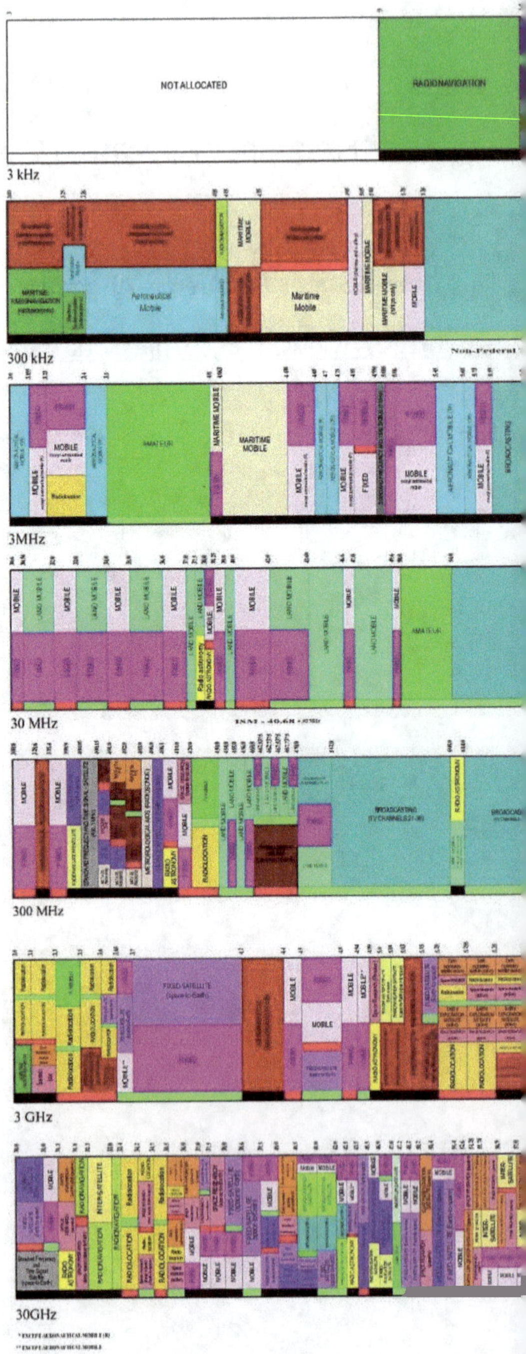

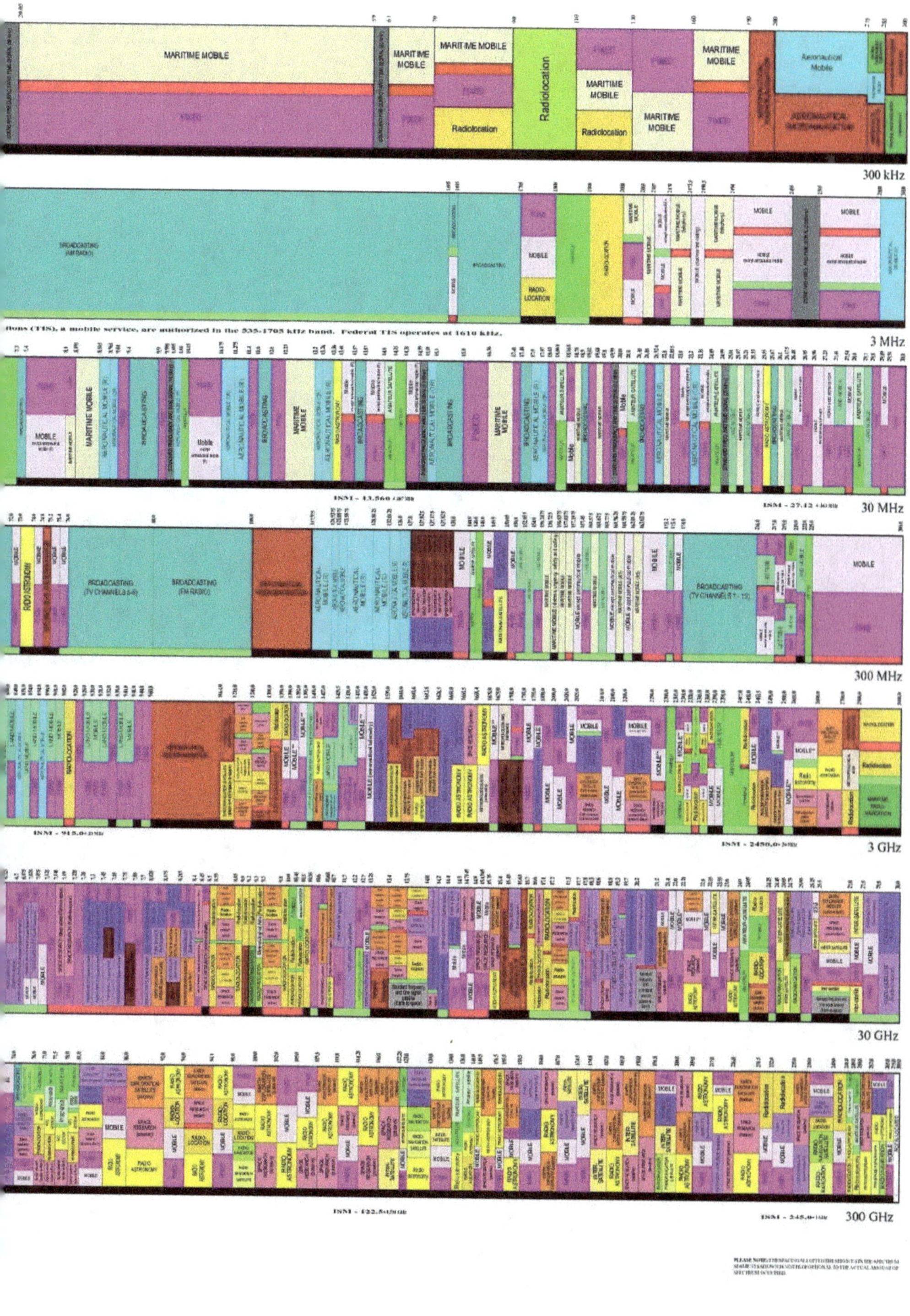
MARITIME MOBILE
MARITIME MOBILE
MARITIME MOBILE
Radiolocation
Radiolocation
MARITIME MOBILE
Radiolocation
MARITIME MOBILE
MARITIME MOBILE
Aeronautical Mobile
300 kHz
BROADCASTING (AM RADIO)
MOBILE
RADIO-LOCATION
MOBILE
MOBILE
3 MHz
MOBILE
MARITIME MOBILE
BROADCASTING
MARITIME MOBILE
BROADCASTING
MARITIME MOBILE
MOBILE
30 MHz
RADIO ASTRONOMY
BROADCASTING (TV CHANNELS 5-6)
BROADCASTING (FM RADIO)
MOBILE
BROADCASTING (TV CHANNELS 7-13)
MOBILE
300 MHz
RADIOLOCATION
MOBILE
RADIO ASTRONOMY
MOBILE
MOBILE
Radiolocation
3 GHz
MOBILE
RADIOLOCATION
30 GHz
MOBILE
RADIO ASTRONOMY
RADIOLOCATION
300 GHz

www.ingramcontent.com/pod-product-compliance
Lightning Source LLC
LaVergne TN
LVHW052225200926
848481LV00001B/216
9798349559600